AF094602

Graph Algorithms and Applications

Graph Algorithms and Applications

Editors

Serafino Cicerone
Gabriele Di Stefano

MDPI • Basel • Beijing • Wuhan • Barcelona • Belgrade • Manchester • Tokyo • Cluj • Tianjin

Editors

Serafino Cicerone
Department of Information
Engineering, Computer Science
and Mathematics
University of L'Aquila
L'Aquila
Italy

Gabriele Di Stefano
Department of Information
Engineering, Computer Science
and Mathematics
University of L'Aquila
L'Aquila
Italy

Editorial Office
MDPI
St. Alban-Anlage 66
4052 Basel, Switzerland

This is a reprint of articles from the Special Issue published online in the open access journal *Algorithms* (ISSN 1999-4893) (available at: www.mdpi.com/journal/algorithms/special_issues/ Graph_Algorithms_Applications).

For citation purposes, cite each article independently as indicated on the article page online and as indicated below:

LastName, A.A.; LastName, B.B.; LastName, C.C. Article Title. *Journal Name* **Year**, *Volume Number*, Page Range.

ISBN 978-3-0365-1542-7 (Hbk)
ISBN 978-3-0365-1541-0 (PDF)

Contents

About the Editors

Serafino Cicerone

Serafino Cicerone received a PhD degree from the University "La Sapienza" of Rome in 1997. He is currently an Associate Professor with the Department of Information Engineering, Computer Science and Mathematics, University of L'Aquila. His research interests revolve around the specification, design, verification and implementation of efficient algorithms. Specific areas of interest include algorithmic graph theory, combinatorial optimization, distributed algorithms, algorithm engineering, and spatial and geometric data.

Gabriele Di Stefano

Gabriele Di Stefano received a PhD degree from the University "La Sapienza" of Rome, in 1992. He is currently a Full Professor with the Department of Information Engineering, Computer Science and Mathematics, University of L'Aquila. He has had key-participations in several EU funded projects. Among them: MILORD (AIM 2024), COLUMBUS (IST 2001-38314), AMORE (HPRN-CT-1999-00104), ARRIVAL (IST FP6-021235-2), and recently GEOSAFE (H2020-691161). His current research interests include algorithmic graph theory, combinatorial optimization, network algorithms, and distributed computing.

Editorial

Special Issue on "Graph Algorithms and Applications"

Serafino Cicerone * and **Gabriele Di Stefano**

Department of Information Engineering, Computer Science and Mathematics, University of L'Aquila, I-67100 L'Aquila, Italy; gabriele.distefano@univaq.it
* Correspondence: serafino.cicerone@univaq.it

Abstract: The mixture of data in real life exhibits structure or connection property in nature. Typical data include biological data, communication network data, image data, etc. Graphs provide a natural way to represent and analyze these types of data and their relationships. For instance, more recently, graphs have found new applications in solving problems for emerging research fields such as social network analysis, design of robust computer network topologies, frequency allocation in wireless networks, and bioinformatics. Unfortunately, the related algorithms usually suffer from high computational complexity, since some of these problems are NP-hard. Therefore, in recent years, many graph models and optimization algorithms have been proposed to achieve a better balance between efficacy and efficiency. The aim of this Special Issue is to provide an opportunity for researchers and engineers from both academia and the industry to publish their latest and original results on graph models, algorithms, and applications to problems in the real world, with a focus on optimization and computational complexity.

Keywords: analysis and design or graph algorithms; distributed graph and network algorithms; graph theory with algorithmic applications; computational complexity of graph problems; experimental evaluation of graph algorithms

Citation: Cicerone, S.; Di Stefano, G. Special Issue on "Graph Algorithms and Applications". *Algorithms* **2021**, *14*, 150. https://doi.org/10.3390/a14050150

Received: 26 April 2021
Accepted: 6 May 2021
Published: 10 May 2021

Publisher's Note: MDPI stays neutral with regard to jurisdictional claims in published maps and institutional affiliations.

1. Introduction

Graphs represent mathematical abstractions that can be used to represent networks of various types: physical (e.g., the Internet or transportation networks), biological (e.g., brain networks), or social (e.g., online social networks). This led the development of algorithmic graph theory as a classical research area in computer science. It focuses on the discovery of characterization theorems on (different types of) graphs, which in turn often lead to the development of efficient algorithms for practical problems that can be modeled on graphs.

2. Special Issue

In response to the call for papers, a total of eighteen manuscripts were submitted. Out of them, we selected six submissions to appear in this Special Issue. In what follows, we summarize the contents of all six published papers.

In [1], the authors faced a typical problem concerning the visual analysis of real-world networks. To this end, they introduce and study the following beyond-planarity problem that they call h-CLIQUE2PATH PLANARITY. Let G be a simple topological graph for which the vertices are partitioned into subsets of size at most h, each inducing a clique: h-CLIQUE2PATH PLANARITY asks whether it is possible to obtain a planar subgraph of G by removing edges from each clique so that the subgraph induced by each subset is a path. They investigate the complexity of this problem in relation to k-planarity. In particular, they prove that h-CLIQUE2PATH PLANARITY is NP-complete even when $h = 4$ and G is a simple 3-plane graph, while it can be solved in linear time when G is a simple 1-plane graph, for any value of h. The results provided contribute to the growing fields of hybrid planarity and of graph drawing beyond planarity.

In [2], the authors used graph theory models to cope with problems arising in the field of molecular biology and bioinformatics. They considered the ancestral mixture model proposed by Chen and Lindsay in 2006, an important model building a hierarchical tree from high dimensional binary sequences. As a phylogenetic tree (or evolutionary tree), a mixture tree created from ancestral mixture models involves the inferred evolutionary relationships among various biological species. Moreover, it contains the information of time when the species mutates. The tree comparison metric, an essential issue in bioinformatics, is used to measure the similarity between trees. Since the approach to the comparison between two mixture trees is still unknown, the authors proposed a new metric to measure the similarity of two mixture trees and designed efficient algorithms for computing it.

In [3], the authors proposed graph models and algorithms for social network analysis. In particular, they considered the phenomenon occurring in many political campaigns where social influence is used in order to convince voters to support/oppose a specific candidate/party. In election control via the social influence problem, an attacker tries to find a set of limited influencers to start disseminating a political message in a social network of voters. A voter changes their opinion when they receive and accept the message. In constructive case, the goal is to maximize the number of votes/winners of a target candidate/party, while in the destructive case, the attacker tries to minimize them. Recent works considered the problem in different models and presented some hardness and approximation results. In that paper, the authors considered multi-winner election control through social influence on different graph structures and diffusion models, and the goal was to maximize/minimize the number of winners in our target party. They showed that the problem is hard to approximate when voters' connections form a graph, and the diffusion model is the linear threshold model. They also proved the same result considering an arborescence under independent cascade model. Moreover, they presented a dynamic programming algorithm for the cases that the voting system is a variation of straight-party voting and voters form a tree.

In [4], the authors considered congestion games, a well-known class of noncooperative games that have the capability to model several interesting competitive scenarios while maintaining nice properties. In these games, there is a set of players sharing a set of resources. Each resource has an associated cost function, which depends on the number of players using it (the so-called congestion). Players aim to choose subsets of resources to minimize the sum of resource costs. In particular, the authors introduced multidimensional congestion games, that is, congestion games for which the set of players is partitioned into $d+1$ clusters $C_0, C_1, \ldots, C_d$. Players in C_0 have full information about all of the other participants in the game, while players in C_i, for any $1 \leq i \leq d$, have full information only about the members of $C_0 \cup C_i$ and are unaware of the others. This model has at least two interesting applications: (*i*) it is a special case of graphical congestion games induced by an undirected social knowledge graph with independence number equal to d, and (*ii*) it represents scenarios in which players have a type and the level of competition they experience on a resource depends on their type and on the types of the other players using it. The authors focused on the case in which the cost function associated with each resource is affine and bound to the price of anarchy and stability as a function of d with respect to two meaningful social cost functions and for both weighted and unweighted players. They also provided refined bounds for the special case of $d = 2$ in the presence of unweighted players.

The remaining two papers addressed typical problems in algorithmic graph theory. In [5], the authors studied the maximum-clique independence problem and some variations of the clique transversal problem such as the $\{k\}$-clique, maximum-clique, minus clique, signed clique, and k-fold clique transversal problems from algorithmic aspects for k-trees, suns, planar graphs, doubly chordal graphs, clique perfect graphs, total graphs, split graphs, line graphs, and dually chordal graphs. They gave equations to compute the $\{k\}$-clique, minus clique, signed clique, and k-fold clique transversal numbers for suns and showed that the $\{k\}$-clique transversal problem is polynomial-time solvable for graphs in which the clique transversal numbers equal their clique independence numbers. They also showed the relationship between the signed and generalization clique problems and presented NP-completeness results for the considered problems on k-trees with unbounded k, planar graphs, doubly chordal graphs, total graphs, split graphs, line graphs, and dually chordal graphs.

Finally, in [6], the class of k-distance-hereditary graphs was studied. The considered graphs have nice properties for which the distance in each connected induced subgraph is at most k times the distance in the whole graph. The defined graphs represent a generalization of the well-known distance-hereditary graphs, which actually correspond to 1-distance-hereditary graphs. This paper provides characterizations for the class of all k-distance-hereditary graphs such that $k < 2$. The new characterizations are given in terms of both forbidden subgraphs and cycle-chord properties. Such results also lead to devising a polynomial-time recognition algorithm for this type of graph that, according to the provided characterizations, simply detects the presence of quasi-holes in any given graph.

Funding: This research received no external funding.

Institutional Review Board Statement: Not applicable.

Informed Consent Statement: Not applicable.

Acknowledgments: The guest editors thank all of the authors who submitted their work to this Special Issue, the reviewers for their constructive comments, and the editorial staff for their assistance.

Conflicts of Interest: The authors declare no conflict of interest.

References

1. Angelini, P.; Eades, P.; Hong, S.H.; Klein, K.; Kobourov, S.; Liotta, G.; Navarra, A.; Tappini, A. Graph Planarity by Replacing Cliques with Paths. *Algorithms* **2020**, *13*, 194. [CrossRef]
2. Juan, J.S.T.; Chen, Y.C.; Lin, C.H.; Chen, S.C. Efficient Approaches to the Mixture Distance Problem. *Algorithms* **2020**, *13*, 314. [CrossRef]
3. Abouei Mehrizi, M.; D'Angelo, G. Multi-Winner Election Control via Social Influence: Hardness and Algorithms for Restricted Cases. *Algorithms* **2020**, *13*, 251. [CrossRef]
4. Bilò, V.; Flammini, M.; Gallotti, V.; Vinci, C. On Multidimensional Congestion Games. *Algorithms* **2020**, *13*, 261. [CrossRef]
5. Lee, C.M. Algorithmic Aspects of Some Variations of Clique Transversal and Clique Independent Sets on Graphs. *Algorithms* **2021**, *14*, 22. [CrossRef]
6. Cicerone, S. A Quasi-Hole Detection Algorithm for Recognizing k-Distance-Hereditary Graphs, with $k < 2$. *Algorithms* **2021**, *14*, 105. [CrossRef]

Article

Graph Planarity by Replacing Cliques with Paths [†]

Patrizio Angelini [1], Peter Eades [2], Seok-Hee Hong [2], Karsten Klein [3], Stephen Kobourov [4], Giuseppe Liotta [5], Alfredo Navarra [6,*] and Alessandra Tappini [5]

[1] School of Computer Science, John Cabot University, 00165 Rome, Italy; pangelini@johncabot.edu
[2] School of Computer Science, Faculty of Engineering, The University of Sydney, Sydney 2006, Australia; peter.edas@sydney.edu.au (P.E.); seokhee.hong@usyd.edu.au (S.-H.H.)
[3] Department of Computer Science, University of Konstanz, 78464 Konstanz, Germany; karsten.klein@uni-konstanz.de
[4] Department of Computer Science, University of Arizona, Tucson, AZ 85721, USA; kobourov@cs.arizona.edu
[5] Department of Engineering, University of Perugia, 06123 Perugia, Italy; giuseppe.liotta@unipg.it (G.L.); alessandra.tappini@unipg.it (A.T.)
[6] Department of Mathematics and Computer Science, University of Perugia, 06123 Perugia, Italy
*   Correspondence: alfredo.navarra@unipg.it; Tel.: +39-075-585-5046
†   This paper is an extended version of our paper published in proceedings of the 26th International Symposium on Graph Drawing and Network Visualization (GD), Barcelona, Spain, 26–28 September 2018. The work began at the Bertinoro Workshop on Graph Drawing (BWGD), Bertinoro, Italy, 4–9 March 2018.

Received: 3 July 2020; Accepted: 11 August 2020; Published: 13 August 2020

Abstract: This paper introduces and studies the following beyond-planarity problem, which we call h-CLIQUE2PATH PLANARITY. Let G be a simple topological graph whose vertices are partitioned into subsets of size at most h, each inducing a clique. h-CLIQUE2PATH PLANARITY asks whether it is possible to obtain a planar subgraph of G by removing edges from each clique so that the subgraph induced by each subset is a path. We investigate the complexity of this problem in relation to k-planarity. In particular, we prove that h-CLIQUE2PATH PLANARITY is NP-complete even when $h = 4$ and G is a simple 3-plane graph, while it can be solved in linear time when G is a simple 1-plane graph, for any value of h. Our results contribute to the growing fields of hybrid planarity and of graph drawing beyond planarity.

Keywords: planar graphs; k-planarity; NP-hardness; polynomial time reduction; cliques; paths

1. Introduction

A typical problem concerning the visual analysis of real-world networks refers to the creation of occlusions and hairball-like structures in dense subnetworks when node-link diagrams are generated by standard layout algorithms, e.g., force-directed methods. On the other hand, different representations, such as adjacency matrices, are well suited for dense graphs but make neighbor identification and path-tracing more difficult [1,2]. *Hybrid graph representations* combine different visualization metaphors in order to exploit their strengths and overcome their drawbacks.

The *NodeTrix* model [3] represents a first example of hybrid representation. It combines node-link diagrams with adjacency-matrix representations of the denser subgraphs [3–6]. Inspired by NodeTrix, other hybrid representation models were recently introduced [7–9]. The *ChordLink* model [7] embeds chord diagrams, used for the visualization of dense subgraphs (*clusters*), into a node-link diagram. In a (k, p) *representation* [8], each cluster contains at most k vertices and each vertex can occur at most p times along the boundary of the cluster. In the *intersection-link representations* [9] model, vertices are geometric objects and edges are either intersections between objects (*intersection-edges*) or crossing-free Jordan arcs attaching at their boundary (*link-edges*). Different types of objects determine different intersection-link representations.

Clique-planar drawings are defined in [9] as intersection-link representations in which the objects are isothetic rectangles, and the partition into intersection- and link-edges is given as a part of the input, so that the graph induced by the intersection-edges is composed of a set of vertex-disjoint cliques. The corresponding recognition problem, called CLIQUE-PLANARITY, has been proved NP-complete in general and polynomial-time solvable in restricted cases, for example when the rectangle representing each vertex is given as a part of the input, or when the cliques are arranged on levels according to a hierarchy. In [9], it is also proven that, if a graph is clique-planar, then it admits an intersection-link representation in which all vertices in a same cluster are isothetic unit squares whose upper-left corners are aligned along a line of slope one (see Figure 1a,b).

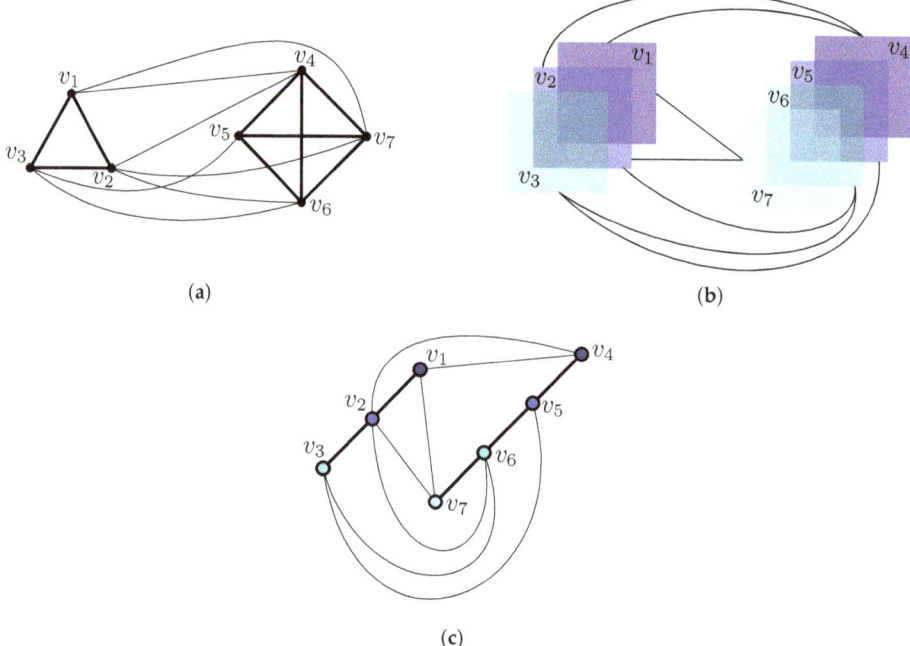

Figure 1. (a) A non-planar graph G. Cliques are highlighted with bold edges. (b) A clique-planar drawing of G. (c) Replacing each clique by a path spanning its vertices. Note that, different from (a), in (c), the first vertex and the last vertex of each path have only one place to connect to edges, while the interior vertices have two places: this is what makes the problem non-trivial.

Therefore, we can reformulate the CLIQUE-PLANARITY problem in the terminology of *beyond-planarity* [10,11] as follows. Given a graph $G = (V, E)$ and a partition of its vertex set V into subsets $V_1, \ldots, V_m$ such that the subgraph of G induced by each subset V_i is a clique, the goal is to compute a planar subgraph $G' = (V, E')$ of G by replacing the clique induced by V_i, for each $i = 1, \ldots, m$, with a path spanning the vertices of V_i (see Figure 1c).

In this paper, we introduce and study a problem called h-CLIQUE2PATH PLANARITY (for short, h-C2PP), that is a restricted version of CLIQUE-PLANARITY in which the input graph comes with a given embedding and each clique has size at most h. Preliminary results have been presented in [12].

1.1. Our Results

A graph G is *planar* if it admits an embedding in the plane where no two edges cross; this embedding is a *planar embedding* of G. A planar graph with an associated planar embedding is said to be an *embedded planar* graph, or a *plane* graph.

In the version of h-CLIQUE2PATH PLANARITY that we study, the input graph G is a *simple topological graph*. A *topological* graph is embedded in the plane so that each edge is a Jordan arc connecting its end-vertices. A topological graph is *simple* if a Jordan arc does not pass through any vertex, and does not intersect any arc more than once (either with a proper crossing or sharing a common end-vertex); finally, no three arcs mutually cross at the same point.

Our main goal is to investigate the complexity of h-C2PP in relation to the well-studied class of *k-planar graphs*, i.e., those that admit a drawing in which each edge has at most k crossings [9,10,13,14]. With a slight abuse of notation, we use the term *embedding* also for non-planar graphs, where we interpret each crossings as a dummy vertex. In particular, a *k-planar* graph together with a *k-planar* embedding is a *k-plane* graph.

A *geometric graph* is drawn in the plane so that each edge is a straight line segment. The version of h-C2PP in which the input graph G is a geometric graph has been recently studied by Kindermann et al. [15], who called it the *partition spanning forest problem*. They proved that 4-C2PP for geometric graphs is NP-complete, which immediately implies the NP-completeness of 4-C2PP for simple topological graphs.

We strengthen this result by proving that 4-C2PP is NP-complete even for simple topological 3-plane graphs. On the positive side, we prove that the h-C2PP problem for simple topological 1-plane graphs can be solved in linear time for any value of h. We finally remark that the 2-SAT formulation used in [15] to solve 3-C2PP for geometric graphs can be easily extended to solve 3-C2PP for any simple topological graph.

1.2. Outline

In Section 2, we further investigate the relationship between h-C2PP and the partition spanning forest problem, that is the problem studied by Kindermann et al. [15]. In Section 3, we prove the NP-completeness of 4-C2PP for simple topological 3-plane graphs. In Section 4, we show that the h-C2PP problem for simple topological 1-plane graphs is linear-time solvable for any value of h. Finally, in Section 5, we provide challenging open problems.

2. Relationship between h-CLIQUE2PAH PLANARITY and the Partition Spanning Forest Problem

The input of the problem studied by Kindermann et al. [15] is a set of colored points in the plane, and the goal is to decide whether there exist straight-line spanning trees, one for each same-colored point subset, that do not cross each other. Since edges are straight-line, their drawings are determined by the positions of the points, and hence each same-colored point subset can, in fact, be seen as a straight-line drawing of a clique, from which edges have to be removed so that each clique becomes a tree and the drawing becomes planar.

The authors proved NP-completeness for the case in which the spanning tree is a path, even when there are at most four vertices with the same color. This result implies that 4-C2PP for geometric graphs is NP-complete. On the other hand, they provided a linear-time algorithm when there exist at most three vertices with the same color, which then extends to 3-C2PP for geometric graphs.

Although not explicitly mentioned in [15], the drawings produced by the reduction used to prove the NP-completeness of 4-C2PP for geometric graphs are 4-planar. We now provide some details about this reduction.

The authors of [15] performed a polynomial-time reduction from PLANAR 3-SATISFIABILITY. The variable gadget (shown in the yellow region of Figure 1) consists of a triangle X whose edges are x, x_l, and x_r. Edge x is crossing-free and the truth value of X is encoded according to which edge among x_l and x_r is crossing-free. Let T_1 and T_2 be two triangles whose vertices are u, y, z and v, y, z, respectively. They define two faces f_1 and f_2, respectively. Concatenate a triangle T_3 defined as in the variable gadget with f_1 by inserting its crossing-free edge (y, z) inside f_1 and by crossing the other two edges of T_3 with (u, y) and (u, z), respectively. Now, concatenate another triangle T_4 defined as in the variable gadget with f_2. If the crossing-free edge of T_4 is inside f_2, the gadget composed by T_1, T_2, T_3 and T_4 is the wire gadget; if the crossing-free edge of T_4 is outside f_2, the gadget composed by T_1, T_2, T_3 and T_4 is the inverter gadget. The splitting gadget consists of three variable gadgets X, Y and Z, and two 4-cliques, concatenated as illustrated inside the blue region in Figure 2, where the yellow region contains a variable gadget, the orange region contains a wire gadget and the violet region contains an inverter gadget. As shown in Figure 2, multiple splittings of a variable X lead to an instance where a triangle has two edges with four crossings.

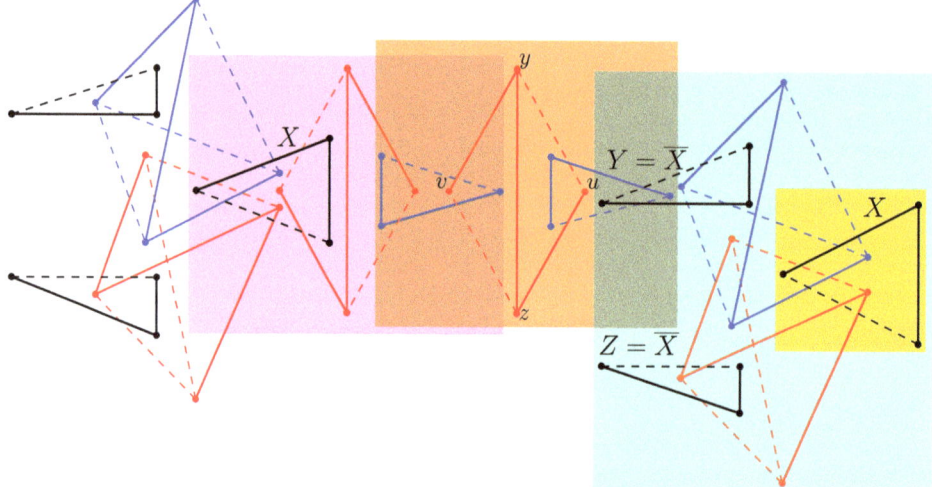

Figure 2. A drawing produced by the reduction in [15]. The yellow region contains a variable gadget, the blue region contains a splitting gadget, the orange region contains a wire gadget, and the violet region contains an inverter gadget.

The NP-completeness of 4-C2PP for geometric graphs implies the NP-completeness of 4-C2PP for simple topological 4-plane graphs. In what follows, we further explore the complexity of 4-C2PP in relation to k-planarity by considering values of $k < 4$. In particular, we prove that the problem remains NP-complete for $k = 3$, while it becomes linear-time solvable for $k = 1$.

3. NP-Completeness for Simple Topological 3-Plane Graphs

In this section, we prove that the 4-C2PP problem remains NP-complete even when the input is a simple topological 3-plane graph.

Since the planarity of a simple topological graph can be checked in linear time, the h-C2PP problem for simple topological k-plane graphs belongs to NP for all values of h and k.

In the following, we prove the NP-hardness by means of a reduction from the PLANAR POSITIVE 1-IN-3-SAT problem. In this version of the SATISFIABILITY problem, which is known to be NP-complete [16], each variable appears only with its positive literal, each clause has at most three variables, the graph obtained by connecting each variable with all the clauses it belongs to is planar, and the goal is to find a truth assignment in such a way that, for each clause, exactly one of its three variables is set to True. Our reduction is technically different from the one presented in [15], which reduces from PLANAR 3-SATISFIABILITY.

For each 3-clique we use in the reduction, there is a *base edge*, which is crossing-free in the constructed topological graph, while the other two edges always have crossings. We call *left* (*right*) the edge that follows (precedes) the base edge in the clockwise order of the edges along the 3-clique. In addition, if an edge e of a clique does not belong to the path replacing the clique, we say that e is *removed*, and that all the crossings involving e in G are *resolved*.

For each variable x, let n_x be the number of clauses containing x. We construct a simple topological graph gadget G_x for x, called *variable gadget* (see the left dotted box in Figure 3a). This gadget contains $2n_x$ 3-cliques $t_1^x, \ldots, t_{2n_x}^x$, forming a ring, so that the left (right) edge of t_i^x only crosses the left (right) edge of t_{i-1}^x and of t_{i+1}^x, for each $i = 1, \ldots, 2n_x$. In addition, gadget G_x contains n_x additional 3-cliques, called $\tau_1^x, \ldots, \tau_{n_x}^x$, so that the right edge of τ_j^x crosses the left edge of t_{2j-1}^x and the right edge of t_{2j}^x, while the left edge of τ_j^x crosses the left edge of t_{2j}^x and the right edge of t_{2j-1}^x.

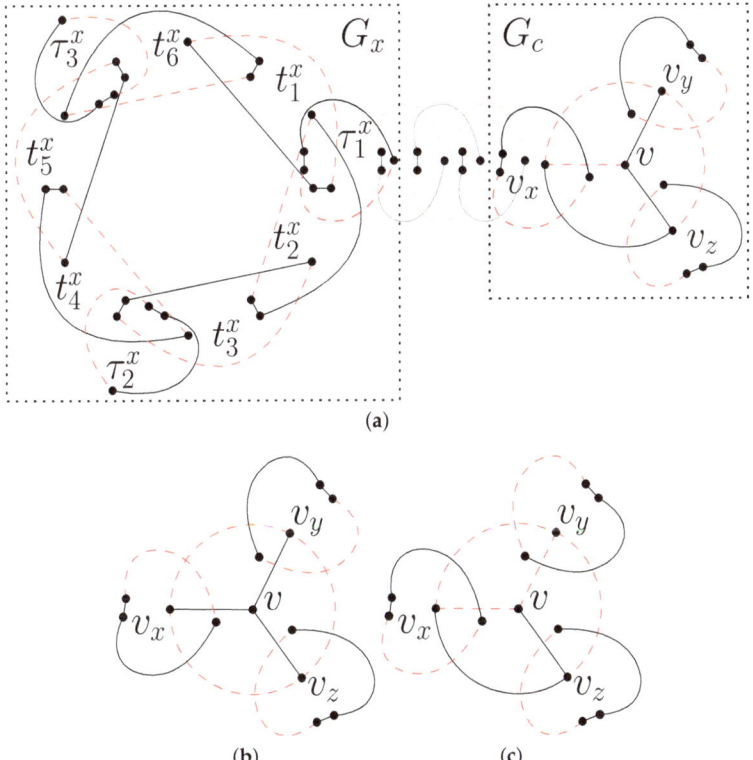

(a)

(b)   (c)

Figure 3. (a) The variable gadget G_x for a variable x is represented in the left dotted box. The clause gadget for a clause c is represented in the right dotted box. The chain connecting G_x to G_c is represented with lighter colors. The removed edges are dashed red. (b) All variables are False. (c) At least two variables are True.

Then, for each clause c, we construct a simple topological graph gadget G_c, called *clause gadget*, which is composed of a planar drawing of a 4-clique, together with three 3-cliques whose left and right edges cross the edges of the 4-clique as in the right dotted box in Figure 3a. In particular, observe that the right (left) edge of each 3-clique crosses exactly one (two) edges of the 4-clique.

Every 3-clique in G_c corresponds to one of the three variables of c. Let x be one of such variables; assuming that c is the jth clause that contains x according to the order of the clauses in the given formula, we connect the 3-clique corresponding to x in the clause gadget G_c to the 3-clique τ_j^x of the variable gadget G_x of x by a chain of 3-cliques of odd length, as in Figure 3a.

By construction, the resulting simple topological graph G contains cliques of size at most 4, namely one per clause, and hence is a valid instance of 4-C2PP. In addition, by collapsing each variable and clause gadget into a vertex, and each chain connecting them into an edge, the resulting graph G' preserves the planarity of the PLANAR POSITIVE 1-IN-3-SAT instance. This implies that the only crossings for each edge of G are with other edges in the gadget it belongs to and, possibly, with the edges of the 3-cliques of a chain. Hence, G is 3-planar. Namely, each base edge is crossing-free; each internal edge of a 4-clique has one crossing; each external edge of a 4-clique has two crossings, and the same is true for the left and right edges of each 3-clique in a chain; finally, the left and right edges of each 3-clique in either a variable or a clause gadget have three crossings.

In the following, we prove the equivalence between the original instance of PLANAR POSITIVE 1-IN-3-SAT and the constructed instance G of 4-C2PP. For this, we first give a lemma stating that variable gadgets correctly represent the behavior of a variable; indeed, they can assume one out of two possible states in any solution for 4-C2PP.

Lemma 1. *Let G_x be the variable gadget for a variable x in G. Then, in any solution for 4-C2PP, either the left edge of each 3-clique τ_j^x, with $j = 1, \ldots, n_x$, is removed, or the right edge of each 3-clique τ_j^x is removed.*

Proof. We first consider the possible removals of edges in $t_1^x, \ldots, t_{2n_x}^x$ and claim that, in any solution for 4-C2PP, one of the two following conditions are satisfied: (i) for each 3-clique t_i^x, if i is odd, then the left edge is removed, while if i is even the right edge is removed; and (ii) for each 3-clique t_i^x, if i is odd, then the right edge is removed, while if i is even the left edge is removed. Note that this claim is sufficient to prove the statement; in fact, if Condition (i) holds (as in Figure 3a), then the right edge of each 3-clique τ_j^x must be removed, in order to resolve its crossings with the right edge of t_{2j-1}^x and with the left edge of t_{2j}^x, while if Condition (ii) holds, then the left edge of each 3-clique τ_j^x must be removed, in order to resolve its crossings with the left edge of t_{2j-1}^x and with the right edge of t_{2j}^x.

To prove the claim, we consider the possible removals of edges of t_1^x. Suppose first that the base edge of t_1^x is removed. Thus, the crossings between the left (right) edge of t_1^x and the left (right) edge of t_2^x are not resolved; this implies that they have to be resolved by removing both the left and the right edge of t_2^x, which is not possible. If the right edge of t_1^x is removed, then the crossing between the right edges of t_1^x and t_2^x is resolved, while the one between their left edges is not. Hence, the left edge of t_2^x must be removed. By iterating this argument we conclude that the right (left) edge of each t_i^x with i odd (even) is removed. Symmetrically, we can prove that, if the left edge of t_1^x is removed, then the left (right) edge of each t_i^x with i odd (even) is removed. This concludes the proof of the lemma. □

Given Lemma 1, we can associate the truth value of a variable x with the fact that either the left or the right edge of each 3-clique τ_j^x in the variable gadget G_x of G is removed. We use this association to prove the following theorem.

Theorem 1. *The 4-C2PP problem is NP-complete, even for 3-plane graphs.*

Proof. Given an instance of PLANAR POSITIVE 1-IN-3-SAT, we construct an instance G of 4-C2PP in linear time as described above. We prove their equivalence.

Suppose first that there exists a solution for 4-C2PP, i.e., a set of edges of G whose removal resolves all crossings. By Lemma 1, for each variable x either the left or the right edge of each 3-clique τ_j^x in the variable gadget G_x is removed. If the right edge is removed, we assign value True to variable x, otherwise we assign False.

To prove that this assignment results in a solution for the given formula of PLANAR POSITIVE 1-IN-3-SAT, we first show that, for each clause c that contains variable x, the right (left) edge of the 3-clique $t_c(x)$ of the clause gadget G_c corresponding to x is removed if and only if the right (left) edge of each 3-clique τ_j^x is removed. Namely, consider the chain that connects $t_c(x)$ with a 3-clique τ_j^x of G_x. Note that, for any two consecutive 3-cliques along the chain, the left edge of one 3-clique and the right edge of the other 3-clique must be removed. Since the chain has odd length, the right (left) edge of $t_c(x)$ is removed if and only if the right (left) edge of τ_j^x is removed, that is, the truth value of G_x is transferred to the 3-clique $t_c(x)$ of G_c.

Finally, consider any clause c, composed of variables x, y, and z. Let $t_c(x)$, $t_c(y)$, and $t_c(z)$ be the three 3-cliques of the clause gadget G_c of c corresponding to x, y, and z, respectively; also, let v be the central vertex of the 4-clique of G_c, and let v_x, v_y, and v_z be the vertices of this 4-clique lying inside $t_c(x)$, $t_c(y)$, and $t_c(z)$, respectively; see Figure 3. We assume without loss of generality that v_x, v_y, and v_z appear in this clockwise order around v. As discussed above, the left or the right edge of $t_c(x)$ (of $t_c(y)$; of $t_c(z)$) is removed depending on whether the left or the right edge of each τ_j^x (of each τ_j^y; of each τ_j^z) is removed. We show that, for exactly one of $t_c(x)$, $t_c(y)$, and $t_c(z)$ the right edge is removed, which then implies that exactly one of x, y, and z is True, and hence the instance of PLANAR POSITIVE 1-IN-3-SAT is positive.

Suppose first that for each of $t_c(x)$, $t_c(y)$, and $t_c(z)$ the left edge is removed (and hence all the three variables are set to False), as in Figure 3b. This implies that the crossings between the right edges of the three 3-cliques and the three edges of triangle (v_x, v_y, v_z) are not resolved. Hence, all the edges of this triangle should be removed, which is not possible since the remaining edges of the 4-clique do not form a path.

Suppose now that for at least two of $t_c(x)$, $t_c(y)$, and $t_c(z)$, say $t_c(x)$ and $t_c(y)$, the right edge is removed (and hence x and y are set to True), as in Figure 3c. Since each edge of triangle (v_x, v_y, v) is crossed by the left edge of one of $t_c(x)$ and $t_c(y)$, by construction, these crossings are not resolved. Hence, all the edges of (v_x, v_y, v) should be removed, which is not possible since the remaining edges of the 4-clique do not form a path of length 4.

Suppose finally that for exactly one of $t_c(x)$, $t_c(y)$, and $t_c(z)$, say $t_c(x)$, the right edge is removed (and hence x is the only one to be set to True), as in Figure 3a. Then, by removing edges (v, v_x), (v_x, v_y), and (v_y, v_z), all the crossings are resolved and the remaining edges of the 4-clique form a path of length 4, as desired.

The proof of the other direction is analogous. Namely, suppose that there exists a truth assignment that assigns a True value to exactly one variable in each clause. Then, for each variable x that is set to True (to False), we remove the right (left) edge of each 3-clique t_i^x, with $i = 2j - 1$ and $j = 1, \ldots, n_x$, we remove the left (right) edge of each 3-clique t_i^x, with $i = 2j$ and $j = 1, \ldots, n_x$, and we remove the right (left) edge of each 3-clique τ_j^x, with $j = 1, \ldots, n_x$. Then, we remove the left or right edge of each 3-clique in a chain so that for any two consecutive 3-cliques, one of them has been removed the left edge and the other one the right edge. This ensures that, for each clause c, the right edge of exactly one of the three 3-cliques that belong to the clause gadget G_c has been removed, say the one corresponding to variable x, while for the other two 3-cliques the left edge has been removed. Hence, we can resolve all crossings by removing edges (v, v_x), (v_x, v_y), and (v_y, v_z), as discussed above (see Figure 3a). The statement follows. □

4. *h*-CLIQUE2PAH PLANARITY and 1-Planarity

In this section, we show that, when the given simple topological graph is 1-plane, *h*-C2PP can be solved in linear time in the size of the input, for any h. We consider all possible simple topological 1-plane cliques and show that the problem can be solved using only local tests, each requiring constant time. Note that we can restrict to the case $h \leq 6$, since K_6 is the largest 1-planar complete graph [11].

Simple topological 1-plane graphs containing cliques with at most four vertices that cross each other can be constructed, but it is easy to enumerate all these graphs (up to symmetry) (see Figure 4). Note that such graphs involve at most two cliques and that, if K_4 has a crossing, combining it with any other clique would violate 1-planarity (see Figure 4a,b). The next lemma accounts for cliques with five or six vertices.

Lemma 2. *There exists no 1-plane simple topological graph that contains two cliques, one of which with at least five vertices, whose edges cross each other.*

Proof. Consider a simple 1-plane graph G that contains two disjoint cliques K and H, with five and three vertices, respectively. Let K' be the simple plane topological graph obtained from K by replacing each crossing with a dummy vertex. By 1-planarity, every face of K' is a triangle and contains at most one dummy vertex. Suppose, for a contradiction, that there exists a crossing between an edge of K and an edge of H in G. Then, there would exist at least a vertex v of H inside a face f of K' and at least one outside f. Since H is a triangle, there must have been two edges that connect vertices inside f to vertices outside f. If f contains one dummy vertex, then two of its edges are not crossed by edges of H, as otherwise G would not be 1-planar. Hence, both the edges that connect vertices inside f to vertices outside f cross the other edge of f, a contradiction. If f contains no dummy vertices, then each edge of f admits one crossing. Let u be the vertex of f that is incident to the two edges crossed by edges of H. Since u has degree 4 in K, it is not possible to draw the third edge of H so that it crosses only one edge of K, which completes the proof. $\square$

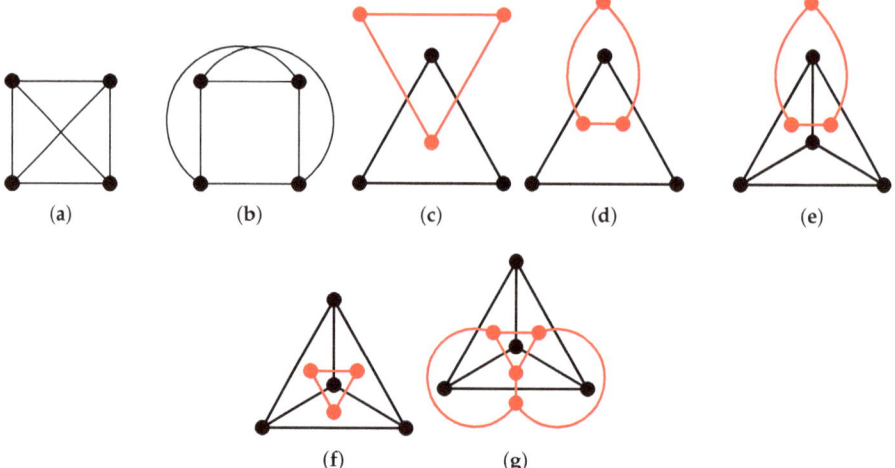

Figure 4. All possible 1-plane graphs involving one or more cliques of type K_3 and K_4 admitting crossings edges. (**a**) and (**b**): two representations of a clique of type K_4; (**c**) and (**d**): two representations of two intersecting cliques of type K_3; (**e**) and (**f**): two representations of a clique of type K_3 intersecting a clique of type K_4; (**g**): two intersecting cliques of type K_4.

Combining the previous discussion with Lemma 2, we conclude that, for each subgraph of the input graph G that consists either of a combination of at most two cliques of size at most 4, as in Figure 4, or of a single clique not crossing any other clique, the crossings involving this subgraph (possibly with other edges not belonging to cliques) can only be resolved by removing its edges, which can be checked in constant time. In the next theorem, n denotes the number of vertices.

Theorem 2. *h-C2PP is $O(n)$-time solvable for simple topological 1-plane graphs.*

5. Conclusions and Open Problems

We introduce and study the h-CLIQUE2PATH PLANARITY problem for simple topological k-plane graphs; we proved that this problem is NP-complete for $h = 4$ and $k = 3$, while it is solvable in linear time for every value of h, when $k = 1$. The natural open question is: What is the complexity for simple topological 2-plane graphs?

Algorithms **2020**, *13*, 194

Kindermann et al. [15] recently proved that problem 4-C2PP is NP-complete for geometric 4-plane graphs. It would be interesting to study this geometric version of the problem for 2-plane and 3-plane graphs.

Recall that the version of the *h*-C2PP problem when the input is an *n*-vertex abstract graph and $h \in O(n)$ is NP-complete, since it is equivalent to CLIQUE PLANARITY [9]. What if the input is an abstract graph and *h* is bounded by a constant or sublinear function? We remark that for $h = 3$ this version of the problem is equivalent to CLUSTERED PLANARITY, when restricted to instances in which the graph induced by each cluster consists of three isolated vertices.

Finally, another intriguing research direction is to study the *h*-CLIQUE2PATH PLANARITY problem in the scenario in which the input graph comes without a clustering of its vertex set, but dense portions of the graph are found by an algorithm. While the problem of finding cliques in a graph is NP-complete [17], one could identify dense subgraphs, for example *k*-cores, in polynomial time [18].

Author Contributions: Conceptualization, All authors; Methodology, All authors; Software, All authors; Validation, All authors; Formal analysis, All authors; Investigation, All authors; Resources, All authors; Data curation, All authors; Writing—original draft preparation, All authors; Writing—review and editing, All authors; Visualization, All authors; Supervision, All authors; Project administration, All authors; Funding acquisition, All authors; All authors have read and agreed to the published version of the manuscript.

Funding: The research was partially supported by: (i) MIUR-DAAD Joint Mobility Program n.57397196 (P.A.); (ii) ARC (Australian Research Council) DP project (S.H.); (iii) Young Scholar Fund/AFF - Univ. Konstanz (K.K.); (iv) NSF grants CCF-1740858 - CCF-1712119 (S.K.); (v) MIUR grant 20174LF3T8 "AHeAD: efficient Algorithms for HArnessing networked Data" (G.L., A.T.); (vi) Dipartimento di Ingegneria dell'Università degli Studi di Perugia, grant RICBA19FM: "Modelli, algoritmi e sistemi per la visualizzazione di grafi e reti" (G.L., A.T.); and (vii) projects "Algorithms and Emergency", "Robot-based computing systems", "Distributed Computing by mobile entities" funded by Fondo Ricerca di Base 2017, 2018, 2019, respectively, University of Perugia (A.N.).

Conflicts of Interest: The authors declare no conflict of interest.

References

1. Ghoniem, M.; Fekete, J.; Castagliola, P. On the readability of graphs using node-link and matrix-based representations: A controlled experiment and statistical analysis. *Inf. Vis.* **2005**, *4*, 114–135. [CrossRef]
2. Okoe, M.; Jianu, R.; Kobourov, S.G. Node-Link or Adjacency Matrices: Old Question, New Insights. *IEEE Trans. Vis. Comput. Graph.* **2019**, *25*, 2940–2952. [CrossRef] [PubMed]
3. Henry, N.; Fekete, J.; McGuffin, M.J. NodeTrix: A Hybrid Visualization of Social Networks. *IEEE Trans. Vis. Comput. Graph.* **2007**, *13*, 1302–1309. [CrossRef] [PubMed]
4. Da Lozzo, G.; Di Battista, G.; Frati, F.; Patrignani, M. Computing NodeTrix Representations of Clustered Graphs. *J. Graph Algorithms Appl.* **2018**, *22*, 139–176. [CrossRef]
5. Di Giacomo, E.; Liotta, G.; Patrignani, M.; Rutter, I.; Tappini, A. NodeTrix Planarity Testing with Small Clusters. *Algorithmica* **2019**, *81*, 3464–3493. [CrossRef]
6. Yang, X.; Shi, L.; Daianu, M.; Tong, H.; Liu, Q.; Thompson, P. Blockwise Human Brain Network Visual Comparison Using NodeTrix Representation. *IEEE Trans. Vis. Comput. Graph.* **2017**, *23*, 181–190. [CrossRef] [PubMed]
7. Angori, L.; Didimo, W.; Montecchiani, F.; Pagliuca, D.; Tappini, A. ChordLink: A New Hybrid Visualization Model. In Proceedings of the Graph Drawing and Network Visualization—27th International Symposium, GD, Prague, Czech Republic, 17–20 September 2019; Volume 11904, pp. 276–290. [CrossRef]
8. Di Giacomo, E.; Lenhart, W.J.; Liotta, G.; Randolph, T.W.; Tappini, A. (k, p)-Planarity: A Relaxation of Hybrid Planarity. In Proceedings of the WALCOM: Algorithms and Computation—13th International Conference, Guwahati, India, 27 February–2 March 2019; pp. 148–159. [CrossRef]
9. Angelini, P.; Da Lozzo, G.; Di Battista, G.; Frati, F.; Patrignani, M.; Rutter, I. Intersection-Link Representations of Graphs. *J. Graph Algorithms Appl.* **2017**, *21*, 731–755. [CrossRef]
10. Didimo, W.; Liotta, G.; Montecchiani, F. A Survey on Graph Drawing Beyond Planarity. *ACM Comput. Surv.* **2019**, *52*, 4:1–4:37. [CrossRef]
11. Kobourov, S.G.; Liotta, G.; Montecchiani, F. An annotated bibliography on 1-planarity. *Comput. Sci. Rev.* **2017**, *25*, 49–67. [CrossRef]

12. Angelini, P.; Eades, P.; Hong, S.; Klein, K.; Kobourov, S.G.; Liotta, G.; Navarra, A.; Tappini, A. Turning Cliques into Paths to Achieve Planarity. In Proceedings of the 26th International Symposium on Graph Drawing and Network Visualization (GD), Barcelona, Spain, 26–28 September 2018; Volume 11282, pp. 67–74.

13. Bekos, M.A.; Kaufmann, M.; Raftopoulou, C.N. On Optimal 2- and 3-Planar Graphs. In Proceedings of the 33rd International Symposium on Computational Geometry, SoCG 2017, Brisbane, Australia, 4–7 July 2017; pp. 16:1–16:16. [CrossRef]

14. Pach, J.; Tóth, G. Graphs Drawn with Few Crossings per Edge. *Combinatorica* **1997**, *17*, 427–439. [CrossRef]

15. Kindermann, P.; Klemz, B.; Rutter, I.; Schnider, P.; Schulz, A. The Partition Spanning Forest Problem. In Proceedings of the 34th European Workshop on Computational Geometry (EuroCG'18), Franconia, Germany, 21–23 March 2018; p. 53.

16. Mulzer, W.; Rote, G. Minimum-weight triangulation is NP-hard. *J. ACM* **2008**, *55*, 1–29. [CrossRef]

17. Karp, R.M. Reducibility Among Combinatorial Problems. In Proceedings of the symposium on the Complexity of Computer Computations, New York, NY, USA, 20–22 March 1972; pp. 85–103. [CrossRef]

18. Batagelj, V.; Zaversnik, M. Fast algorithms for determining (generalized) core groups in social networks. *Adv. Data Anal. Classif.* **2011**, *5*, 129–145. [CrossRef]

Article

Efficient Approaches to the Mixture Distance Problem

Justie Su-Tzu Juan [1], Yi-Ching Chen [2], Chen-Hui Lin [1] and Shu-Chuan Chen [3,*]

1   Department of Computer Science and Information Engineering, National Chi Nan University,
     Puli, Nantou 54561, Taiwan; jsjuan@ncnu.edu.tw (J.S.-T.J.); tedlinct@gmail.com (C.-H.L.)
2   Department of Computer Science and Information Engineering, National Taiwan University,
     Taipei 10617, Taiwan; d94010@csie.ntu.edu.tw
3   Department of Mathematics and Statistics, Idaho State University, Pocatello, ID 83209, USA
*   Correspondence: scchen@isu.edu

Received: 31 August 2020; Accepted: 23 November 2020; Published: 28 November 2020

Abstract: The ancestral mixture model, an important model building a hierarchical tree from high dimensional binary sequences, was proposed by Chen and Lindsay in 2006. As a phylogenetic tree (or evolutionary tree), a mixture tree created from ancestral mixture models, involves the inferred evolutionary relationships among various biological species. Moreover, it contains the information of time when the species mutates. The tree comparison metric, an essential issue in bioinformatics, is used to measure the similarity between trees. To our knowledge, however, the approach to the comparison between two mixture trees is still unknown. In this paper, we propose a new metric named the mixture distance metric, to measure the similarity of two mixture trees. It uniquely considers the factor of evolutionary times between trees. If we convert the mixture tree that contains the information of mutation time of each internal node into a weighted tree, the mixture distance metric is very close to the weighted path difference distance metric. Since the converted mixture tree forms a special weighted tree, we were able to design a more efficient algorithm to calculate this new metric. Therefore, we developed two algorithms to compute the mixture distance between two mixture trees. One requires $O(n^2)$ and the other requires $O(nh_1h_2)$ computational time with $O(n)$ preprocessing time, where n denotes the number of leaves in the two mixture trees, and h_1 and h_2 denote the heights of these two trees.

Keywords: phylogenetic tree; evolutionary tree; ancestral mixture model; mixture tree; mixture distance; tree comparison

1. Introduction

Phylogeny reconstruction involves reconstructing the evolutionary relationship from biological sequences among species. Nowadays it has become a critical issue in molecular biology and bioinformatics. Several existing methods, such as neighbor-joining methods [1] and maximum likelihood methods [2], have been proposed to reconstruct a phylogenetic tree. A novel and natural method, ancestral mixture models [3], was developed by Chen and Lindsay to deal with such a problem. The mixture tree, a hierarchical tree created from the ancestral mixture model, induces a sieve parameter to represent the evolutionary time. Chen, Rosenberg and Lindsay (2011) then developed MixtureTree algorithm [4], a linux based program written in C++, which employed the ancestral mixture models to reconstruct mixture tree from DNA sequences. With the information provided by the mixture tree, one can identify when and how a mutation event of species occurs. An example of the mixture tree created by MixtureTree algorithm [3] is shown in Figure 1. The data from Griffiths and Tavare (1994) [5] are a subset of the mitochondrial DNA sequences which first appeared in Ward et al. (1991) [6]. To study the mitochondrial diversity within the Nuu-Chuah-Nulth, an Amerindian tribe from Vancouver Island, Ward et al. (1991) [6] sequenced 360 nucleotide segments of the mitochondrial control region for 63 individuals from the Nuu-Chuah-Nulth. Griffiths' and Tavares' subsample consisted of 55 of the 63 distinct sequences and 18 segregating sites, including 13 pyrimidines (C, T) and five purines (A, G). Each linage represents a distinct sequence—that is, there are lineages *a* through *n*. The time scale on the tree can be represented by $-\log(1-2p)$, where *p* is a parameter, the mutation rate. The number on the tree represents the site of the lineage whereat the mutation occurs. For example, when $p = 0.01$, lineages *e* and *f* merge because mutation occurs at site 5 of lineage *f*.

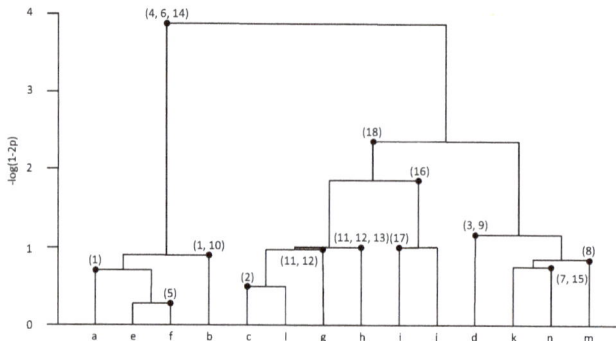

Figure 1. An example of the mixture tree [3].

Distinct methods may produce distinct trees, even though the methods adopt an identical dataset [7]. To uncover a well-represented tree involved in evolutionary relationship among species it is quite important to estimate how similar (or different) trees are. The tree distance between two trees is a general measurement for the similarity of the trees.

The tree distance problem is a traditional issue in mathematics. Several metrics have been proposed to measure the similarity between two trees, such as the partition metric (also called the Robison–Foulds metric or RF distance for short) [8], the quartet metric [9], the nearest neighbour interchange metric [10] and the nodal distance metric [11]. Those metrics all compare two trees by considering the tree structure only, and do not mention any parameter in the tree. Thus, those metrics are not suitable for computing the similarity between two mixture trees. Therefore, we propose a novel metric named the mixture distance metric to measure the similarity of two mixture trees in this paper. Among the above metrics, the metric from the nodal distance algorithm is similar to our proposed metric. In 2003, John Bluis and Dong-Guk Shin [11] presented the nodal distance algorithm which is used to measure the distances from leaves to all other leaves in a tree. The metric is defined as follows: $\text{Distance}(T_1, T_2) = \sum_{x,y \in L(T_1) = L(T_2)} |D_{T_1}(x,y) - D_{T_2}(x,y)|$, where $D_{T_i}(x,y)$ denotes the distance of leaf *x* to leaf *y* in the tree T_i. The nodal distance algorithm was developed for this metric. Anyway, using this metric to measure the distance between two mixture trees is not conformable.

For the metric of the mixture distance, the time parameter indicating when a mutation event of species occurs plays an important role in the tree similarity, which is, however, not considered by those previous metrics. If the weight of an edge in a mixture tree is defined as the difference in time parameters between its two endpoints, a mixture tree can be regarded as a weighted tree. We can design metrics to calculate the distance between two weighted trees. Some literature discusses the distance problem between two weighted trees. For example, take the weighted RF metric [12], geodesic distance [13] and the path difference metric [14]. However, the weight on each edge is considered to be the number of base changes between the sequences of the species represented by its incident vertices in these documents. Since the weights of each edge in those weighted trees may be different, the algorithm must spend more time to calculate those distances between those two weighted trees. For example, although there is an linear time algorithm to compute RF distance [15], and a randomized algorithm has been shown to approximate the RF distance with a bounded error in sublinear time [16], the complexity of the weighted RF distance still needs $O(n^2)$. Some papers have studied algorithms for calculating the geodesic trees distances [17–19]. The best one already known is $O(n^4)$ [19]. Due to the characteristics of the time parameter of a mixture tree, any two edges connecting two leaves to the same parent will have the same "weight" in a mixture tree. This helped us to design a better metric and algorithm. We further developed two algorithms to compute the mixture distance between two mixture trees. One requires $O(n^2)$ and the other requires $O(nh_1h_2)$ computational time with $O(n)$ preprocessing time, where n denotes the number of leaves in these two mixture trees, and h_1 and h_2 denote the heights of these two trees. If we use the nodal distance algorithm with the mixture distance metric, the time complexity will be $O(n^3)$ for binary unrooted trees. Comparisons with some previous methods show our method performs better.

2. Mixture Distance Metric

A tree $T = (V(T), E(T))$ is a connected and acyclic graph with a node set $V(T)$ and an edge set $E(T)$. T is a rooted tree if exactly one node of T has been designated the root. A node $v \in V(T)$ is a leaf if it has no child; otherwise, v is an internal node. A node $v \in V(T)$ is called in level i, denoted by $level(v) = i$, which means the number of edges on the path between the root and v is i. Let $L(T)$ denote a subset of node set $V(T)$, where each member is a leaf in T and $n = |L(T)|$. Let $height(T)$ denote the height of tree T, which is $\max\{level(v) | v \in L(T)\}$. T is a full binary tree if each node of T either has two children or it is a leaf. A complete binary tree is a full binary tree in which every level, except possibly the last, is completely filled, and all nodes are as far left as possible. Let $h_1 = height(T_1)$, $h_2 = height(T_2)$.

For a mixture tree T, each leaf is associated with a species, and every internal node v is associated with a mutation time $m_T(v)$ that represents the time when a mutation event occurs on the species node. In fact, the mutation time of an internal node in a mixture tree can be regarded as the distance between the node and any leaf of its descendants. Any two mixture tress T_1 and T_2 are comparable if $L(T_1) = L(T_2)$. Throughout this paper, a tree refers to a rooted full binary tree and each internal node of the tree is associated with its mutation time, if not mentioned particularly.

Given any two nodes $u, v \in V(T)$, the least common ancestor or lowest common ancestor (abbreviated LCA) of u and v is an ancestor of both u and v with the smallest mutation time. (It is also called the most recent common ancestor (abbreviated MRCA), or the last common ancestor (abbreviated LCA) in biology and genealogy.) Let $P_T(u, v)$ denote the mutation time $m_T(w)$ of the LCA w of two leaves u and v in T. The mixture distance metric, a metric for the mixture tree, is formally defined as follows.

The mixture distance between two comparable mixture trees T_1 and T_2, denoted by $d_m(T_1, T_2)$, is defined as the sum of difference of the mutation times with respect to the LCAs of any two leaves in T_1 and T_2. That is, $d_m(T_1, T_2) = \sum_{u,v \in L(T_1) = L(T_2)} |P_{T_1}(u, v) - P_{T_2}(u, v)|$.

The significance of the mixture distance metric is to measure the similarity between two mixture trees, considering the mutation times (molecular clock) and mutation sites simultaneously. The study sought to develop two algorithms for efficiently computing the mixture distance between two comparable mixture trees. Before we go into the algorithms, three properties of the mixture distance matric are demonstrated. Felsenstein [20] derived three mathematical properties—reflexivity, symmetry and triangle inequality—required for a well-defined metric. We show that the mixture distance is well-defined in Theorem 1.

Theorem 1. *The mixture distance d_m satisfies:*

1. *Reflexivity: for any two comparable mixture trees T_1 and T_2, $d_m(T_1, T_2) = 0$ if and only if T_1 and T_2 are identical.*
2. *Symmetry: for any two comparable mixture trees T_1 and T_2, $d_m(T_1, T_2) = d_m(T_2, T_1)$.*

3. *Triangle inequality: for any three comparable mixture trees T_1, T_2 and T_3, $d_m(T_1, T_2) + d_m(T_2, T_3) \geq d_m(T_1, T_3)$.*

Proof. 1. Due to $T_1 = T_2$, for any two nodes $u, v \in L(T_1) = L(T_2)$, we have $P_{T_1}(u, v) = P_{T_2}(u, v)$. Therefore, $d_m(T_1, T_2) = 0$ can be concluded. On the other hand, if $d_m(T_1, T_2) = 0$ for any two comparable mixture trees T_1 and T_2. We have $P_{T_1}(u, v) - P_{T_2}(u, v)$ for any $u, v \in L(T_1) = L(T_2)$ by the definition. Then we can prove $T_1 = T_2$ by induction on the height of T_1 (or T_2).

2. For any two nodes $u, v \in L(T_1) = L(T_2)$, $P_{T_1}(u, v) - P_{T_2}(u, v) = -(P_{T_2}(u, v) - P_{T_1}(u, v))$. Thus, $d_m(T_1, T_2) = \sum_{u,v \in L(T_1)=L(T_2)} |P_{T_1}(u, v) - P_{T_2}(u, v)| = \sum_{u,v \in L(T_1)=L(T_2)} |P_{T_2}(u, v) - P_{T_1}(u, v)| = d_m(T_2, T_1)$.

3. The triangle inequality is always satisfied for any three nonnegative numbers $a, b, c \in \Re^+ \cup 0$; that is, $|a - b| + |b - c| \geq |a - c|$. Therefore, $|P_{T_1}(u, v) - P_{T_2}(u, v)| + |P_{T_2}(u, v) - P_{T_3}(u, v)| \geq |P_{T_1}(u, v) - P_{T_3}(u, v)|$ holds. Further, we have

$$\sum_{u,v \in L(T_1)} |P_{T_1}(u, v) - P_{T_2}(u, v)| + \sum_{u,v \in L(T_2)} |P_{T_2}(u, v) - P_{T_3}(u, v)|$$

$$\geq \sum_{u,v \in L(T_1)} |P_{T_1}(u, v) - P_{T_3}(u, v)|.$$

Consequently, $d_m(T_1, T_2) + d_m(T_2, T_3) \geq d_m(T_1, T_3)$ can be concluded. □

3. An $O(nh_1h_2)$-Time Algorithm

Let T_1 and T_2 denote two comparable mixture trees of n leaves for each tree. Note that the mixture distance of T_1 and T_2 can be solved in $O(n^2)$-time: As when given two comparable mixture trees T_1 and T_2 each with n leaves, there are $O(n^2)$ pairs of leaves separately in T_1 and T_2. In fact, the LCA of any pair of leaves can be found by adopting the $O(1)$-time algorithm with $O(n)$-time preprocessing [21].

In the following, another $O(n^2)$-time algorithm, named Algorithm MIXTUREDISTANCE, is proposed to compute the mixture distance between T_1 and T_2, which will help us to realize the next $O(nh_1h_2)$-time algorithm, the main result.

3.1. Algorithm MixtureDistance

Algorithm MIXTUREDISTANCE, as shown on Algorithm 1, proceeds the nodes of T_1 by breadth-first search. For each internal node v in T_1, we find out the leaves of T_1 such that v is exactly the LCA of each pair of leaves, and then compute the LCA u of the leaves in T_2 which are mapped into the found leaves of T_1. Finally, the difference of the mutation times between u and v is calculated. For convenience, we define $(a, b) * (c, d) = ad + bc$ for any two ordered pairs (a, b) and (c, d) in this algorithm, where a, b, c and d are any four integers.

The algorithm adopts a 2-coloring method [22] on the leaves in T_1 and T_2 for easy implementation. For each iteration associated with an internal node v of T_1 in line 4, the leaves of the left and right subtrees rooted by v are colored by red and green, respectively. The mapped leaves in T_2 have the same coloring as one in T_1. The mixture distance between each internal node u in T_2 and v is calculated according the coloring scheme in T_2 (in lines 16–17), and the coloring information of u would be derived for the computation of its parent node (in line 18).

The coloring information of u, denoted by $color(u)$, indicates the coloring information of the subtree in T_2 rooted by u. $color(u)$ includes two numbers of u's descendant leaves colored by red ($color(u)[0]$) and green ($color(u)[1]$), respectively. $color(u)$ is derived by the coloring information of its two children. That is, $color(u)[0] = color(u_L)[0] + color(u_R)[0]$ and $color(u)[1] = color(u_L)[1] + color(u_R)[1]$, where u_L and u_R separately denote the left and right children of u in T_2.

Algorithm 1: MIXTUREDISTANCE(T_1, T_2).

Input: Two comparable mixture trees T_1 and T_2, with mutation times $m_{T_1}(v)$ ($m_{T_2}(u)$, respectively) for every internal node v of T_1 (u of T_2, respectively).

Output: The mixture distance $\mathcal{D}$ between T_1 and T_2.

1   $\mathcal{D} = 0$.

2   Traverse T_1 by the breadth-first search from its root and keep a list $\mathcal{I}_1$ of the internal nodes in order.

3   Traverse T_2 by the breadth-first search from its root and keep a list $\mathcal{I}_2$ of the internal nodes in reverse order.

4   **for** each node $v \in \mathcal{I}_1$ **do**

5   In T_1, color red the leaves of the left subtree rooted by v and green the leaves of the right subtree rooted by v.

6   **for** each node $u \in \mathcal{I}_2$ **do**

        // Initialize the coloring information of u's children

7   **for** each child w of u in T_2 **do**

8   **if** w is a leaf **then**

9   **if** w is colored by red in T_1 **then**

10              $color(w) = (1, 0)$.

11   **else if** w is colored by green in T_1 **then**

12              $color(w) = (0, 1)$.

13   **else**

14              $color(w) = (0, 0)$.

15   Let u_L and u_R be the left and right children of u in T_2, respectively.

        // Calculate the difference of the mutation times of u and v and sum them up for computing mixture distance

16         $number(u) = color(u_L) * color(u_R)$.

17         $\mathcal{D} = \mathcal{D} + |m_{T_1}(v) - m_{T_2}(u)| \times number(u)$.

        // Calculate the coloring information of u

18         $color(u) = color(u_L) + color(u_R)$.

In line 16, $number(u)$ is achieved by the special product of the color vectors of u's two children, $number(u) = color(u_L)[0] \times color(u_R)[1] + color(u_L)[1] \times color(u_R)[0]$, which means the number of times that u is an LCA of a red leaf and a green leaf. We multiply the difference of their mutation times by $number(u)$ in line 17, for computing the mixture distance between each internal node u in T_2 and v. At the end of Algorithm MIXTUREDISTANCE, $\mathcal{D}$ indicates the mixture distance of T_1 and T_2.

Since the numbers of internal nodes in T_1 and T_2 (= $\mathcal{I}_1$ and $\mathcal{I}_2$) are both equal to $n - 1$, two for-loops will take $O(n)$ time, and the innermost for-loop always takes 2 (a constant) time units. Therefore, Algorithm MIXTUREDISTANCE requires $O(n^2)$ computational time.

3.2. Modified Algorithm

After introducing Algorithm MIXTUREDISTANCE, we can give a $O(nh_1h_2)$ computational time algorithm for computing the mixture distance between two mixture trees in the following part. In Algorithm MIXTUREDISTANCE, when the leaves of the subtree rooted by an internal node v in T_1 are colored, other leaves in T_1 have no color, as do the mapped leaves in T_2. That is, $color(w) = (0, 0)$ for $w \in L(T_2)$. However, Algorithm MIXTUREDISTANCE still processes the ancestors of such leaves in T_2. In the following, we propose an algorithm for disregarding the nodes without meaningful coloring information, and reduce the time complexity from $O(n^2)$ to $O(nh_1h_2)$.

The algorithm contains three main stages, as follows:

1. Rank the leaves in T_1 and T_2.
2. Construct a minimal subtree T_2' of T_2 involved in colored leaves with respect to node v, for each internal node v in T_1.
3. Compute the mixture distance between v and each internal node in T_2'.

In stage 1, the nodes of T_2 are ranked in postorder, and the leaves of T_1 are assigned by the same rank of the mapped leaves in T_2. In Figure 2, red numbers nearby leaves in two given comparable mixture trees T_1 and T_2 indicate the ranking achieved by stage 1 of the algorithm. Note that the number within the nodes means the mutation time $m_{T_i}(v)$ of the associated node v for $i = 1$ or 2.

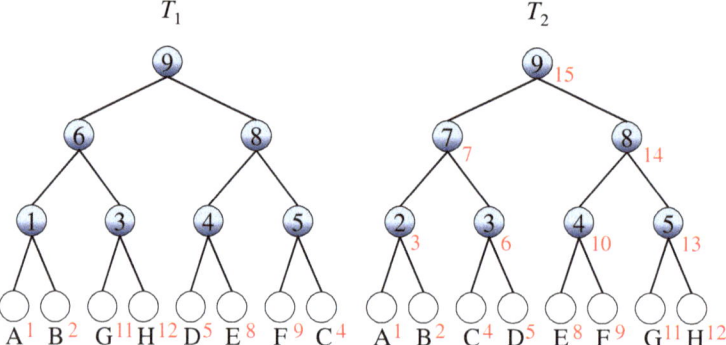

Figure 2. An example of ranking leaves of T_1 and T_2.

The algorithm proceeds to stage 2 for each internal node v of T_1 in the reverse order of breadth-first search. When v in T_1 is processed, stage 2 seeks to construct a minimal subtree T_2' of T_2 involved in colored leaves with respect to node v. For node v, a nondecreasing list of the leaves of the subtree rooted by v, denoted by $leaf(v)$, is obtained from the leaf lists of its two children, where the leaves in the list are sorted by their ranks. Suppose that there are k ordered nodes in $leaf(v)$, that is, $leaf(v) = \{w_1, w_2, \ldots, w_k\}$. With the list $leaf(v)$, the subtree T_2' can be constructed as follows.

Let $lca(w_i, w_j)$ denote the LCA of leaves w_i and w_j in T_2, for any $i, j \in \{1, 2, \ldots, k\}$. The subtree $T_2' = (V', E')$ is initialized by $V' = \{w_1, w_2, lca(w_1, w_2)\}$, $E' = \{\overline{lca(w_1, w_2)w_1}, \overline{lca(w_1, w_2)w_2}\}$ and $root = lca(w_1, w_2)$. For node w_i, $i \in \{1, 2, \ldots, k - 2\}$,

$$V' = V' \cup \{lca(w_{i+1}, w_{i+2}), w_{i+2}\} \text{ and}$$

$$E' = E' \cup \{\overline{lca(w_{i+1}, w_{i+2})w_{i+2}}\}$$

Moreover, if the mutation time (the number written in the node circle) of $lca(w_{i+1}, w_{i+2})$, denoted by $t(lca(w_{i+1}, w_{i+2}))$, is larger than the mutation time of $root$, denoted by $t(root)$, the edge $\overline{lca(w_{i+1}, w_{i+2})root}$ is inserted into E' and reset $lca(w_{i+1}, w_{i+2})$ as the new $root$. Otherwise, if $t(lca(w_{i+1}, w_{i+2}))$ is smaller than the mutation time of $lca(w_i, w_{i+1})$, denoted by $t(lca(w_i, w_{i+1}))$, the edge $\overline{lca(w_i, w_{i+1})w_{i+1}}$ is removed from E' and the edges $\overline{lca(w_{i+1}, w_{i+2})w_{i+1}}$ and $\overline{lca(w_i, w_{i+1})lca(w_{i+1}, w_{i+2})}$ are inserted into E'. Otherwise, let $x = w_{i+1}$ and repeat do $x = father(x)$ until $t(x) < t(lca(w_{i+1}, w_{i+2})) < t(father(x))$, where $father(x)$ is the node y such that $\overline{yx} \in E'$. Then the edge $\overline{father(x)x}$ is removed from E' and the edges $\overline{lca(w_{i+1}, w_{i+2})x}$ and $\overline{father(x)lca(w_{i+1}, w_{i+2})}$ are inserted into E'.

Example 1. *An example of constructing the subtree T_2' with respect to $leaf(v_2) = \{A, B, G, H\}$ is illustrated in Figure 3. Initially, the node set V' is $\{A, B, lca(A, B)\}$ and the edge set E' includes the incident edges of the three nodes in T_2. As node A is processed, two nodes $lca(B, G)$ and G are inserted into V', and two edges $\overline{lca(A,B)lca(B,G)}$ and $\overline{lca(B,G)G}$ are inserted into E'. Later, when node B is processed, two nodes $lca(G, H)$ and H are inserted into V' and two edges $\overline{lca(B,G)lca(G,H)}$ and $\overline{lca(G,H)H}$ are inserted into E'. Meanwhile, the edge $\overline{lca(B,G)G}$ is removed from E' and the edge $\overline{lca(G,H)G}$ is inserted into E', because the mutation time of $lca(B, G)$ is larger than the time of the $lca(G, H)$.*

$$leaf(v_2) = \{A, B, G, H\}$$

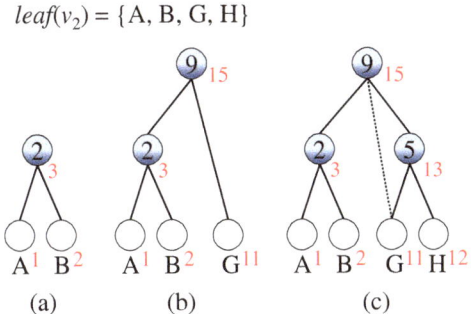

Figure 3. An example of constructing the subtree T_2' with respect to $leaf(v_2)$ in Figure 2. (**a**) The initialization of T_2'. (**b**) The intermediate of T_2' as node A is processed. (**c**) The complete subtree T_2' as node B is processed. As the mutation time of lca(B, G) is larger than the time of lca(G, H), the dotted line incident to G is removed and the other incident edge of G is inserted.

After the subtree T_2' with respect to currently processed node v is constructed, stage 3 of the algorithm performs lines 5–18 of Algorithm MIXTUREDISTANCE to compute the "partial" mixture distance between T_2' and the subtree rooted by v (only computes the distances of some nodes pairs, for which LCA is equal to v). At the end of the algorithm, $\mathcal{D}$ indicates the mixture distance between T_1 and T_2.

Theorem 2. *The improved algorithm takes $O(nh_1h_2)$ computational time and $O(n)$ preprocessing time, where n denotes the number of leaves of the mixture trees and $h_i = height(T_i)$ for $i = 1, 2$.*

Proof. The algorithm contains three main stages. The first stage ranks the leaves in T_1 and T_2, which takes $O(n)$ time.

In the second stage, a minimal subtree T_2' of T_2 involved in colored leaves with respect to each node v in T_1 is constructed. For each node v, a leaf list $leaf(v)$ is obtained from the leaf lists of its two children, which is achieved in $O(t)$ time by using the two-way merging algorithm [23] performed on the leaf list of v's children, where t is the size of $leaf(v)$. The $O(1)$-time algorithm with $O(n)$-time processing [21] is employed to compute the LCA of any pair of nodes in T_2. Constructing T_2' takes $O(th')$ time, where h' is the height of T_2' due to the "repeat" step. The last stage computes the mixture distance between v and each internal node in T_2' by performing lines 5–18 of Algorithm MIXTUREDISTANCE, which takes $O(t)$ time. Stages 2 and 3 take $O(n)$ iterations in total. However, each iteration deals with different t nodes. Note that for all internal nodes which are in the same level of T_1, the sum of t (for each node) is n. Therefore, stages 2 and 3 totally take $O(nh_1h') = O(h_1h_2)$ time, where h_1 is the height of T_1 (note that $h' \le h_2 = height(T_2)$). Hence, the algorithm requires $O(nh_1h_2)$ computational time with $O(n)$ preprocessing time. □

4. Conclusions

In this paper, we provide a novel metric named the mixture distance metric to measure the similarity between two mixture trees. It uniquely considers the estimated evolutionary time in the trees. Two algorithms were developed to compute the mixture distance between mixture trees. One requires $O(n^2)$ computational time and the other requires $O(nh_1h_2)$ computational time with $O(n)$ preprocessing time, respectively. Note that when T_1 and T_2 are complete binary trees, h_1 and h_2 will be $O(\log n)$ and the time complexity of our algorithm will be $(n \log^2 n)$.

Now, we compare our metric with some previous methods which measure phylogenetic differences in consideration of the branch length, when we consider a mixture tree as a weighted tree (recall that the weight of an edge in a mixture tree is defined as the difference of time parameters between its two endpoints). For the geodesic tree distance, the implementation is quite complex and requires heavy computation [19], although a heuristic fast version exists [18]. The definition of the weighted path difference distance [14] is almost the same as the mixture distance. Actually, the weighted path difference distance between two mixture trees T_1 and T_2 is equal to $2d_m(T_1, T_2)$. However, it requires $O(n^2)$ computational time. The mixture distance seems to be similar to the weighted RF distance [12], but the calculation performance will vary when we consider the distance between two different extents of similar mixture trees. We give an example as follows.

Example 2. *Four mixture trees with the same lineages A, B and C are illustrated in Figure 4; the time parameters are listed in the vertices, and the associated edge weights are labeled beside each edge. All pairs of these four trees have been compared using the methods outlined in [12] and this paper. The tables of the weighted RF (wRF) and mixture distances (d_m) are given in Tables 1 and 2, respectively. From these two tables, one can find something interesting. (1) d_m seems maintain the order relationship in wRF: When wRF thinks that two trees are similar, then d_m also gets a smaller value between these two trees: $wRF(T_1, T_3) > wRF(T_2, T_3) > wRF(T_2, T_4) > wRF(T_1, T_2)$ and $d_m(T_1, T_3) > d_m(T_2, T_3) > d_m(T_2, T_4) > d_m(T_1, T_2)$. (2) When wRF thinks that the distances between two pairs of trees are the same, then d_m also thinks they are in the same: $wRF(T_1, T_2) = wRF(T_3, T_4), wRF(T_1, T_4) = wRF(T_2, T_3)$ and $d_m(T_1, T_2) = d_m(T_3, T_4), d_m(T_1, T_4) = d_m(T_2, T_3)$. However, there are still differences between these two metrics in the details: (3) When wRF thinks two distances between two pairs of trees are very different, sometimes d_m may not think that: $wRF(T_1, T_3) - wRF(T_2, T_3) = 1$, $wRF(T_1, T_4) - wRF(T_2, T_4) = 3$, but $d_m(T_1, T_3) - d_m(T_2, T_3) = d_m(T_1, T_4) - d_m(T_2, T_4) = 1$.*

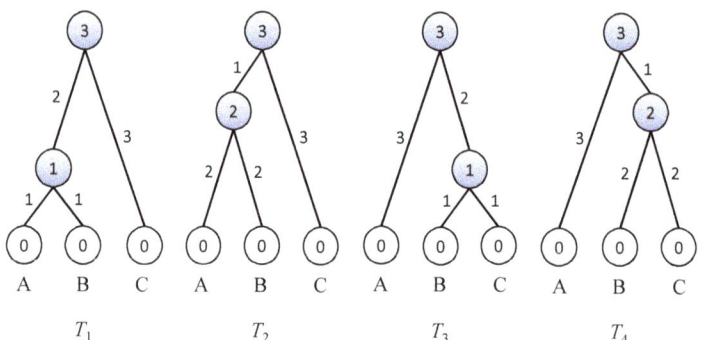

Figure 4. Four weighted trees with the same lineages A, B and C.

Table 1. The weighted RF distances wRF among T_1, T_2, T_3 and T_4.

wRF	T_1	T_2	T_3	T_4
T_1		3	8	7
T_2			7	4
T_3				3

Table 2. The mixture distances d_m among T_1, T_2, T_3 and T_4.

d_m	T_1	T_2	T_3	T_4
T_1		1	4	3
T_2			3	2
T_3				1

Therefore, it can be said that the performance of the mixture distance in calculating the similarity of two weighted trees is as good as the performance of the weighted RF distance, while the time complexity of the mixture distance is better. In addition, we compared our approaches with the methods performed with the nodal distance metric [11], geodesic tree distance [19], weighted path difference metric [14] and weighted RF distance [12], and the results are shown in Table 3. Our proposed approaches performed better than all of the previous methods when discussing the distance between two mixture trees.

Table 3. Comparison of metrics for binary trees.

Metric	Consideration	Time Complexity	
		Full Binary Trees	**Complete Binary Trees**
Mixture distance	Structure and mutation time	$O(nh_1h_2)$	$O(n \log^2 n)$
Nodal distance	Structure	$O(n^3)$	$O(n^2 \log n)$
Geodesic tree distance	Structure and mutation number	$O(n^4)$	$O(n^4)$
Weighted path difference distance	Structure and mutation number	$O(n^2)$	$O(n^2)$
Weighted RF distance	Structure and mutation number	$O(n^2)$	$O(n^2)$

Author Contributions: Investigation, Y.-C.C. and C.-H.L.; methodology, J.S.-T.J. and S.-C.C. All authors have read and agreed to the published version of the manuscript.

Funding: The first author was supported in part by the Ministry of Science and Technology of the Republic of China under Contract No. MOST100-2221-E-260-024- and MOST109-2115-M-260-001.

Conflicts of Interest: The authors declare no conflict of interest.

References

1. Saitou, N.; Nei, M. The neighbor-joining method: A new method for reconstructing phylogenetic trees. *Mol. Biol. Evol.* **1987**, *4*, 406–425. [PubMed]
2. Lesperance, M.L.; Kalbfleisch, J.D. An algorithm for computing the nonparametric MLE of a mixing distribution. *J. Am. Stat. Assoc.* **1992**, *87*, 120–126. [CrossRef]
3. Chen, S.C.; Lindsay, B.G. Building mixture trees from binary sequence data. *Biometrika* **2006**, *93*, 843–860. [CrossRef]
4. Chen, S.C.; Rosenberg, M.; Lindsay, B.G. MixtureTree: A program for constructing phylogeny. *BMC Bioinform.* **2011**, *12*, 111–114. [CrossRef] [PubMed]
5. Griffiths, R.C.; Tavare, S. Ancestral inference in population genetics. *Statist. Sci.* **1994**, *9*, 307–319. [CrossRef]
6. Ward, R.H.; Frazier, B.L.; Dew-Jager, K.; Paabo, S. Extensive mitochondrial diversity within a single amerindian tribe. *Proc. Nat. Acad. Sci. USA* **1991**, *88*, 6720–6724. [CrossRef] [PubMed]
7. Steel, M.A. The maximum likelihood point for a phylogenetic tree is not unique. *Syst. Biol.* **1994**, *43*, 560–564. [CrossRef]
8. Robinson, D.F.; Foulds, L.R. Comparison of phylogenetic trees. *Biosciences* **1981**, *53*, 131–147. [CrossRef]
9. Estabrook, G.F.; McMorris, F.R.; Meacham, C.A. Comparison of undirected phylogenetic trees based on subtrees of four evolutionary units. *Syst. Zool.* **1985**, *34*, 193–200. [CrossRef]
10. Dasgupta, B.; He, X.; Jiang, T.; Li, M.; Tromp, J.; Zhang, L. On computing the nearest neighbor interchange distance. In Proceedings of the Discrete Mathematical Problems with Medical Applications: DIMACS Workshop on Discrete Problems with Medical Applications, Piscataway, NJ, USA, 8–10 December 1999; DIMACS Series in Discrete Mathematics and Theoretical Computer Science; American Mathematical Society: Washington, DC, USA, 2000; Volume 55, pp. 125–143.

11. Bluis, J.; Shin, D. Nodal distance algorithm: calculating a phylogenetic tree comparison metric. In Proceedings of the Third IEEE Symposium on BioInformatics and BioEngineering, Bethesda, MD, USA, 12 March 2003; pp. 87–94.
12. Robinson, D.F.; Foulds, L.R. Comparison of weighted labelled trees. In *Combinatorial Mathematics VI*; Springer: Berlin/Heidelberg, Germany, 1979; pp. 119–126.
13. Billera, L.J.; Holmes, S.P.; Vogtmann, K. Geometry of the space of phylogenetic trees. *Adv. Appl. Math.* **2001**, *27*, 733–767. [CrossRef]
14. Steel, M.A.; Penny, D. Distributions of tree comparison metrics—Some new results. *Syst. Biol.* **1993**, *42*, 126–141.
15. Day, W.H. Optimal algorithms for comparing trees with labeled leaves. *J. Classif.* **1985**, *2*, 7–28. [CrossRef]
16. Pattengale, N.D.; Gottlieb, E.J.; Moret, B.M. Efficiently computing the Robinson-Foulds metric. *J. Comput. Biol.* **2007**, *14*, 724–735. [CrossRef] [PubMed]
17. Battagliero, S.; Puglia, G.; Vicario, S.; Rubino, F.; Scioscia, G.; Leo, P. An efficient algorithm for approximating geodesic distances in tree space. *IEEE/ACM Trans. Comput. Biol. Bioinform.* **2010**, *8*, 1196–1207. [CrossRef] [PubMed]
18. Amenta, N.; Godwin, M.; Postarnakevich, N.; John, K.S. Approximating geodesic tree distance. *Inf. Process. Lett.* **2007**, *103*, 61–65. [CrossRef]
19. Owen, M.; Provan, J.S. A fast algorithm for computing geodesic distances in tree space. *IEEE/ACM Trans. Comput. Biol. Bioinform.* **2010**, *8*, 2–13. [CrossRef] [PubMed]
20. Felsenstein, J. *Inferring Phylogenies*; Sinauer Associates: Sunderland, MA, USA, 2004.
21. Bender, M.A.; Farach-Colton, M. The LCA problem revisited. *Lat. Am. Theor. Inform.* **2000**, *1776*, 88–94.
22. Brodal, G.S.; Fagerberg, R.; Pedersen, C.N.S. Computing the quartet distance between evolutionary trees in time $O(n \log n)$. *Algorithmica* **2003**, *38*, 377–395. [CrossRef]
23. Lee, R.C.T.; Chang, R.C.; Tseng, S.S.; Tsai, Y.T. *Introduction to the Design and Analysis of Algorithms*; McGraw-Hill Education: Berkshire, UK, 2005.

Publisher's Note: MDPI stays neutral with regard to jurisdictional claims in published maps and institutional affiliations.

 *algorithms*

Article

Multi-Winner Election Control via Social Influence: Hardness and Algorithms for Restricted Cases

Mohammad Abouei Mehrizi * and Gianlorenzo D'Angelo *

Computer Science Department, Gran Sasso Science Institute (GSSI), Viale Francesco Crispi,
L'Aquila AQ 67100, Italy
*   Correspondence: mohammad.aboueimehrizi@gssi.it (M.A.M.); gianlorenzo.dangelo@gssi.it (G.D.)

check for updates

Received: 4 September 2020; Accepted: 29 September 2020; Published: 2 October 2020

Abstract: Nowadays, many political campaigns are using social influence in order to convince voters to support/oppose a specific candidate/party. In election control via social influence problem, an attacker tries to find a set of limited influencers to start disseminating a political message in a social network of voters. A voter will change his opinion when he receives and accepts the message. In constructive case, the goal is to maximize the number of votes/winners of a target candidate/party, while in destructive case, the attacker tries to minimize them. Recent works considered the problem in different models and presented some hardness and approximation results. In this work, we consider multi-winner election control through social influence on different graph structures and diffusion models, and our goal is to maximize/minimize the number of winners in our target party. We show that the problem is hard to approximate when voters' connections form a graph, and the diffusion model is the linear threshold model. We also prove the same result considering an arborescence under independent cascade model. Moreover, we present a dynamic programming algorithm for the cases that the voting system is a variation of straight-party voting, and voters form a tree.

Keywords: computational social choice; election control; multi-winner election; social influence; influence maximization

1. Introduction

Social media is an integral part of nowadays life. No one can ignore the effect of social media on different aspects of our life. Many people from all around the world are using social networks to provide/use various services like teaching/learning, spreading information, events' announcements, and advertising. It has been shown that two-thirds of American adults get news on social mediaSM [1]. It is easy to find evidence that a social influence (SI) started by few users has influenced many people. Then, social media is a kind of cheap means to spread a message among many users. Note that the power of social media is not just like spreading a message or advertising. Its power comes from the fact that a user will receive news from those who have enough authority to change his opinion, like close friends, family members, and colleagues. Since using social influence is effective and cheap, it has been attracting the attention of many political campaigns and candidates to target the user's opinion through SI. They disseminate a piece of information to change voters' opinion. Many real case studies show that campaigns used social influence to change the voters' opinion [2–5]. For example, Allcott and Gentzkow showed that 92% of Americans remembered pro-Trump false news, and 23% remembered pro-Clinton fake news [6].

There are two well-known diffusion models used in social influence called linear threshold model (LTM) and Independent Cascade Model (ICM) [7]. In LTM, a voter accepts a message if the sum over his incoming neighbors' influence, who already accepted the message, is high enough. On the other hand, in ICM, a voter will accept a message if at least one of his incoming neighbors, who already accepted the message, can convince him to accept it (please see Section 2 for a formal definition of LTM and ICM).

In this paper, we consider the multi-winner election control via social influence problem. We are given a social network of voters, a limited budget, a set of candidates each belongs to a party, a dynamic diffusion model to spread a message among the voters, and an attacker/manipulator who supports/opposes a party. When we use LT diffusion model, we assume that the attacker knows the probability that each voter wants to vote for each candidate. To take into account the incoming influence of each node v, we use an updating rule based on the incoming influence from the node's incoming activated neighbors, akin to [8]. On the other hand, when we use ICM, we assume the attacker knows the exact preferences list of all voters. When a node/voter becomes active/influenced/infected, in constructive (resp. destructive) case, it will promote (resp. demote) the position of the target candidates in its/his preference list, akin to [9,10] (see Section 3 for formal definition).

Regarding both LTM and ICM, there will be several winners, and they will be elected according to the overall candidates' scores after the diffusion. In the constructive (resp. destructive) case, the attacker wants to find a set of nodes (voters), according to its budget, to start the diffusion and change the voters' opinion to maximize (resp. minimize) the number of winners from his target party. In fact, in a given directed graph, we should find some diffusion starters to influence the voters such that the difference between the number of winners from our target party, w.r.t. the number of winners in the opponent party with the most winners, after and before the diffusion is maximized (resp. minimized). We present some results, including hardness of approximation, approximation, and polynomial-time exact algorithms considering some well-known objective functions on different structures.

Related works. There are many articles regarding voting manipulation (see the survey in [11]). The problem of finding a set of limited seed nodes from a given graph to maximize the expected number of influenced nodes is known as Influence Maximization (IM) problem. There exists an extensive literature about it, too [12]. Domingos and Richardson [13,14] introduced the IM problem, and Kempe et al. formalized it [7,15]. On the other hand, few works consider both of them together, i.e., the election control through social influence problem.

Wilder and Vorobeychik introduced the election control through SI problem regarding single-winner elections [10]. They investigated maximizing margin of victory (MoV) and probability of victory (PoV), where MoV is the difference of the score between the target candidate and the most voted opponent after and before the diffusion. The problem is considered under ICM. They showed maximizing MoV is *NP*-hard, and presented a $1 - \frac{1}{e}$-approximation algorithm concerning the optimal solution. Furthermore, for maximizing PoV, they showed that it is *NP*-hard to approximate the problem within any constant factor. Corò et al. [16,17] extended the work using any non-increasing scoring function under LTM. They demonstrated the same approximation factor for it. Abouei Mehrizi et al. considered the problem when the attacker knows a probability distribution over the candidates instead of the exact preferences list, under LTM [8]. They showed that maximizing/minimizing the expected probability to vote for a target candidate is hard to approximate within any constant factor under unique game with small set expansion conjecture. They also presented some constant factor approximation algorithms for a relaxed version of the problem. Abouei Mehrizi and D'Angelo showed that in multi-winner elections, when the manipulator wants to maximize/minimize the number of winners in his target party, the problem is inapproximable under ICM, except $P = NP$ [9]. They also presented some constant factor approximation algorithms when the voting system is similar to the straight-party voting.

Bredereck and Elkind considered some different models, like bribing nodes/voters, adding or deleting edges under LTM. They showed that the problem is hard in those models. They also presented some polynomial-time algorithms for specific cases of the problem [18]. Castiglioni et al. investigated similar models under ICM. They showed that the problem is hard even in restricted structures. Regarding the bribing nodes to influence other voters, they proved that the election control is hard even if the given graph is a line. Furthermore, considering the edge removal/addition case, they demonstrated that the problem is hard even if the attacker has an infinite budget [19]. Faliszewsk et al. considered the problem where each voter has a preference list. Each node of the graph is representative of all users with the same opinions. There is an edge between two nodes if their opinion differs by the place of an adjacent pair of candidates. They used LTM and proved that maximizing the number of votes for the target candidate is *NP*-hard and fixed parameter tractable with respect to the number of candidates [20]. Furthermore, there is another model in which voters have a preference list over candidates, and voters will change their preference list according to the majority of their neighbors' opinions [21–23].

Outline and our results. In Section 2, we define the most prominent diffusion models in the literature (called LTM and ICM) that we used in this paper. Section 3 defines our model and objective functions formally. We show that our problem is hard to approximate within any factor in a general graph when the diffusion model is LTM in Section 4. Section 5 contains the same result when the diffusion model is ICM, and the given graph is in the form of an arborescence, i.e., edges are from leaves to root of the tree. Moreover, in Section 6, we investigate the problem while the voting system is a variation of straight-party voting, where voters can vote for the parties. In other words, voters have a preference list (or probability distribution) over the candidates, but they can vote for the parties instead of candidates. We presented a polynomial-time algorithm based on the dynamic programming approach to find the maximum difference of votes for our target party before and after diffusion. It also gives a $\frac{1}{3}$ and $\frac{1}{2}$-approximation algorithms for maximizing MoV in constructive and destructive models, respectively. Finally, we will discuss the results and future works in Section 7.

2. Background

In this section, we introduce two diffusion models that we have used in this paper, called linear threshold model (LTM) and independent cascade model (ICM) presented by Kemp et al. [7,15]. They are the most prominent dynamic diffusion models used in literature (see a survey on the topic [24]).

2.1. Linear Threshold Model

We are given a directed graph $G = (V, E)$. Each edge $(u, v) \in E$ has a weight $b_{u,v} \in [0, 1]$. The sum of the incoming weight to each node $v \in V$ is at most one, i.e., $\sum_{u \in N_v^i} b_{u,v} \leqslant 1$, where N_v^i is the set of incoming neighbors of v. Furthermore, each node $v \subset V$ has a threshold $t_v \in [0, 1]$ which is generated uniformly at random.

In this model, the diffusion will start from a set of nodes $S \subseteq V$ known as seed nodes. At the first step, just the seed nodes will become active/influenced/infected, and all other nodes are inactive. Let us show A_i as the set of nodes that are active at step i, i.e., $A_1 = S$. The activation process, for each step $i > 1$, is as follows: all nodes in A_{i-1} will remain active at step i, i.e., $A_{i-1} \subseteq A_i$; moreover, each inactive node $v \in V \setminus A_{i-1}$ will become active if the sum of the weight from its incoming activated neighbors is not less than its threshold, i.e., for each node $v \in V \setminus A_{i-1}$, it will be in A_i if $\sum_{u \in N_v^i} b_{u,v} \geqslant t_v$. The diffusion process will proceed in utmost $|V|$ discrete steps, and it will stop as soon as no extra node becomes active, i.e., it stops at step $k > 1$ if $A_k = A_{k-1}$. We use A_S as the set of activated nodes after the diffusion process started from the set of seed nodes S. In what follows, to increase the readability of this article, when we say after S, it means after the diffusion process started from a set of seed nodes S. Note that the thresholds are not a part of the input, and they will be generated uniformly at random and independently when we run the process. Furthermore, the process is random, and several executions on the same graph may get different results for A_S.

Kemp et al. [7] defined the IM problem as: Given a graph $G = (V, E)$ and a budget $B \leqslant |V|$. Find a set of seed nodes $S \subseteq V$, ($|S| \leqslant B$) so that the expected $|A_S|$ is maximized. They proved that the problem is *NP*-hard under LTM. Moreover, they showed that a greedy algorithm can solve the problem approximately within a factor of $1 - \frac{1}{e} - \epsilon$, where ϵ is any small constant and fixed number.

2.2. Independent Cascade Model

Consider a graph $G = (V, E)$ with a weight $b_{u,v} \in [0, 1]$ on each edge $(u, v) \in E$. The same as LTM, all nodes are inactive, and at the first step the seed nodes $S \subseteq V$ become active. Let us define S_i as the nodes that were inactive at step $i - 1$ and became active at step i, then $S_1 = S$. At each step $i > 1$, each node $v \in S_{i-1}$ will try to activate its outgoing neighbors with the probability of the edge between them. In other words, consider N_v^o as the set of outgoing neighbors of node v; for each $u \in N_v^o$, node v tries to activate u with the probability $b_{v,u}$. If v has multiple outgoing neighbors, it tries to activate them in an arbitrary order. Note that a node becomes active once, let us say at step k, and try to activate its outgoing neighbors exactly once, at step $k + 1$.

Kemp et al. [7] considered the IM under ICM. They showed that the greedy algorithm works for this model, too. They also demonstrated that it is *NP*-hard to approximate the problem within any factor better than $1 - \frac{1}{e}$.

3. Multi-Winner Election Control: Models and Objective Functions

In this section, we consider the Multi-Winner Election Control, where some parties are running for an election so that more than one candidate will be elected as the winner, like a parliament election. We consider t different parties $C_1, \ldots, C_t$, each of them contains k different candidates, i.e., $C_i = \{c_1^i, \ldots, c_k^i\}, 1 \leqslant i \leqslant t$. We use C for the set of all candidates, i.e., $C = \cup_{i=1}^{t} C_i$. Furthermore, without loss of generality, we assume C_1 is our target party. Note that there will be exactly k winners for the election.

3.1. Multi-Winner Election Control under LTM

In this model, we investigate the case that the adversary does not know the preferences list of the voters; instead of that, for each voter, the attacker has a probability distribution over all candidates. This model is similar to the model known as probabilistic linear threshold ranking (PLTR) defined in [8]. Since most voters do not reveal their preferences in social media, then it is a realistic assumption.

The adversary tries to maximize/minimize the number of winners in his target party. For each node $v \in V$, we show π_v as the probability distribution of the voter/node v over all candidates; we define $\pi_v(c)$ as the probability that the voter v votes for a specific candidate $c \in C$. Then for every node $v \in V$, and candidate $c \in C$ we have $\pi_v(c) \in [0, 1]$, and $\sum_{c \in C} \pi_v(c) = 1$.

In LTM, each node has an incoming influence, which shows the amount of pressure from incoming neighbors to support/oppose a target party. We use this incoming influence of node $v \in V$ to change its probability distribution. Let us define $\tilde{\pi}_v$ as the probability distribution of node v after S. Respectively, $\tilde{\pi}_v(c)$ is the probability that node v will vote for candidate $c \in C$ after S. We use A_S to show the set of nodes that will become active after S.

We consider a single message which spreads among the voters. The message contains some constructive/destructive information targeting all candidates in the target party. When a node v becomes active, its probability distribution will change according to the incoming influence from its activated neighbors. We have to normalize the vector in order to make sure that the sum of the probabilities is equal to one, after S. For constructive model the probability distribution of a node $v \in A_S$ changes as follows:

$$\forall c \in C_1 : \tilde{\pi}_v(c) = \frac{\pi_v(c) + \frac{1}{|C_1|} \sum_{u \in A_S \cap N_v^i} b_{uv}}{1 + \sum_{u \in A_S \cap N_v^i} b_{uv}},$$

$$\forall c \in C \setminus C_1 : \tilde{\pi}_v(c) = \frac{\pi_v(c)}{1 + \sum_{u \in A_S \cap N_v^i} b_{uv}}.$$

Recall that N_v^i is the set of incoming neighbors of node v. Furthermore, considering the destructive case, the probability distribution of an active node $v \in A_S$ will change as follows.

$$\forall c \in C_1 : \tilde{\pi}_v(c) = \frac{\pi_v(c)}{1 + \sum_{u \in A_S \cap N_v^i} b_{uv}}$$

$$\forall c \in C \setminus C_1 : \tilde{\pi}_v(c) = \frac{\pi_v(c) + \frac{1}{|C \setminus C_1|} \sum_{u \in A_S \cap N_v^i} b_{uv}}{1 + \sum_{u \in A_S \cap N_v^i} b_{uv}}$$

By these changes (and normalization), we guarantee that the sum of the probability for each node is equal to 1. In both constructive and destructive cases, the probability distribution of inactive nodes $v \in V \setminus A_S$ will not change after S, i.e., $\tilde{\pi}_v = \pi_v$.

Let us define the expected number of votes for candidate $c \in C$ after S, as $\mathcal{F}(c, S) = \mathbb{E}_{A_S}[\sum_{v \in V} \tilde{\pi}_v(c)]$; similarly, $\mathcal{F}(c, \varnothing) = \mathbb{E}[\sum_{v \in V} \pi_v(c)]$ is the expected number of votes for candidate $c \in C$ before any diffusion.

Example 1. *Assume there are two parties supporting two candidates each, i.e., $C = C_1 \cup C_2$, $C_1 = \{c_1^1, c_2^1\}, C_2 = \{c_1^2, c_2^2\}$. There are five nodes in the given graph $G = (V, E)$, where their connections form a star and the weight of all edges is one, i.e., $(v_1, v_2), (v_1, v_3), (v_1, v_4), (v_1, v_5) \in E$, $b_{v_1, v_2} = b_{v_1, v_3} = b_{v_1, v_4} = b_{v_1, v_5} = 1$. Let us consider the probability distribution of each node $v \in V$ as $\pi_v = \pi_v(c_1^1), \pi_v(c_2^1), \pi_v(c_1^2), \pi_v(c_2^2)$. We set the probability distribution of all nodes as $\frac{1}{8}, \frac{1}{8}, \frac{3}{8}, \frac{3}{8}$. Then before any diffusion, the candidates' score is*

$$\mathcal{F}(c_1^1, \varnothing) = \mathcal{F}(c_2^1, \varnothing) = \frac{5}{8},$$

$$\mathcal{F}(c_1^2, \varnothing) = \mathcal{F}(c_2^2, \varnothing) = \frac{15}{8},$$

and none of our target candidates have less score than their opponents. Consider the constructive model in which the adversary's budget is one, i.e., he can select one node to influence the voters and change their opinion. Since the node $v_1 \in V$ is the most influential node in the graph, the adversary selects it as his seed node. It activates all nodes in the graph, and their probability distribution will be updated as follows.

$$\tilde{\pi}_{v_1} = \frac{1}{8}, \frac{1}{8}, \frac{3}{8}, \frac{3}{8},$$

$$\tilde{\pi}_{v_2} = \tilde{\pi}_{v_3} = \tilde{\pi}_{v_4} = \tilde{\pi}_{v_5} = \frac{5}{16}, \frac{5}{16}, \frac{3}{16}, \frac{3}{16},$$

and the expected number of votes for the candidates is

$$\mathcal{F}(c_1^1, S) = \mathcal{F}(c_2^1, S) = \frac{11}{8},$$

$$\mathcal{F}(c_1^2, S) = \mathcal{F}(c_2^2, S) = \frac{9}{8},$$

and our target candidates' score is more than their opponents' score.

3.2. Multi-Winner Election Control under ICM

Our model is similar to the work presented in [9]. We briefly mention the model bellow. In this model, despite LTM, we assume that the attacker knows the voters' preference list. Each voter $v \in V$ has a preferences list π_v. Abusing the notations, $1 \leqslant \pi_v(c) \leqslant tk$ is the rank of candidate c in the preference list of the voter v. After the diffusion, inactive voters will keep their original opinions, i.e., $\forall v \in V \setminus A_S : \tilde{\pi}_v = \pi_v$; however, the activated voters will change their preferences list as follows. Remind that A_S is the set of activated nodes after S.

- Constructive: For each node $v \in A_S$ and for each target candidate $c \in C_1$, the new position of c in $\tilde{\pi}_v$ is

$$\tilde{\pi}_v(c) = \begin{cases} \pi_v(c) - 1 & \text{if } \exists\, c' \in C \setminus C_1 \text{ s.t. } \pi_v(c') < \pi_v(c) \\ \pi_v(c) & \text{otherwise,} \end{cases}$$

also, for other candidates $c \in C \setminus C_1$, if there is a candidate $c' \in C \setminus C_1$ s.t. $\pi_v(c') = \pi_v(c) + 1$, then we set $\tilde{\pi}_v(c) = \pi_v(c)$; otherwise the new rank of the candidate c will be calculated as follows.

$$\tilde{\pi}_v(c) = \pi_v(c) + |\{c'' \in C_1 \mid \pi_v(c'') > \pi_v(c) \land (\nexists\, \bar{c} \in C \setminus C_1 : \pi_v(c) < \pi_v(\bar{c}) < \pi_v(c''))\}|\,.$$

- Destructive: For each node $v \in A_S$ and for each target candidate $c \in C_1$, we have

$$\tilde{\pi}_v(c) = \begin{cases} \pi_v(c) + 1 & \text{if } \exists\, c' \in C \setminus C_1 \text{ s.t. } \pi_v(c') > \pi_v(c) \\ \pi_v(c) & \text{otherwise,} \end{cases}$$

while for $c \in C \setminus C_1$, if there exists a candidate $c' \in C \setminus C_1$ s.t. $\pi_v(c') = \pi_v(c) - 1$ we set $\tilde{\pi}_v(c) = \pi_v(c)$, otherwise we have

$$\tilde{\pi}_v(c) = \pi_v(c) - |\{c'' \in C_1 \mid \pi_v(c'') < \pi_v(c) \land (\nexists\, \bar{c} \in C \setminus C_1 : \pi_v(c'') < \pi_v(\bar{c}) < \pi_v(c))\}|\,.$$

In this article, we consider the plurality scoring rule for simplicity, where just the most preferred candidate of each voter gets one score. However, the results can be extended for any non-increasing scoring function, e.g., k-approval, anti-plurality, and Borda's rule [25]. Let us denote by $\mathcal{F}(c, \varnothing)$, $\mathcal{F}(c, S)$, the expected score of candidate c before and after S, respectively; formally, $\forall c \in C : \mathcal{F}(c, \varnothing) = \sum_{v \in V} \mathbb{1}_{\pi_v(c)=1}$, $\mathcal{F}(c, S) = \mathbb{E}_{A_S}\left[\sum_{v \in V} \mathbb{1}_{\tilde{\pi}_v(c)=1}\right]$. (If we want to generalize the problem and consider any non-increasing scoring function $g(\cdot)$, the functions would be defined as $\mathcal{F}(c, \varnothing) = \sum_{v \in V} g(\pi_v(c))$, $\mathcal{F}(c, S) = \mathbb{E}_{A_S}\left[\sum_{v \in V} g(\tilde{\pi}_v(c))\right]$).

Example 2. *Consider the graph G and candidates C in Example* 1. *Let set the voters' preference list as follows.*

$$\pi_{v_1} = c_1^1 \succ c_2^1 \succ c_1^2 \succ c_2^2,$$
$$\pi_{v_2} = c_1^2 \succ c_2^1 \succ c_1^1 \succ c_2^2,$$
$$\pi_{v_3} = c_2^2 \succ c_1^2 \succ c_1^1 \succ c_2^1,$$
$$\pi_{v_4} = c_1^2 \succ c_2^1 \succ c_1^1 \succ c_2^2,$$
$$\pi_{v_5} = c_2^2 \succ c_1^1 \succ c_1^2 \succ c_2^1,$$

where $a \succ b$ means a is preferred to b. The candidates' score before any diffusion is

$$\mathcal{F}(c_1^1, \varnothing) = 1,$$
$$\mathcal{F}(c_2^1, \varnothing) = 0,$$
$$\mathcal{F}(c_1^2, \varnothing) = \mathcal{F}(c_2^2, \varnothing) = 2,$$

and before any diffusion, both of our target candidates have less score than their opponents. Consider the constructive case where the adversary's budget is one. The same as Example 1, *the adversary selects the node v_1 as a seed node, and it activates all nodes in the graph. After S, the voters update their preference list as follows.*

$$\pi_{v_1} = c_1^1 \succ c_2^1 \succ c_1^2 \succ c_2^2,$$
$$\pi_{v_2} = c_2^1 \succ c_1^1 \succ c_1^2 \succ c_2^2,$$
$$\pi_{v_3} = c_2^2 \succ c_1^1 \succ c_2^1 \succ c_1^2,$$

$$\pi_{v_4} = c_2^1 \succ c_1^1 \succ c_1^2 \succ c_2^2,$$
$$\pi_{v_5} = c_1^1 \succ c_2^2 \succ c_2^1 \succ c_1^2,$$

and the candidates' score will be as follows.

$$\mathcal{F}(c_1^1, S) = \mathcal{F}(c_2^1, S) = 2,$$
$$\mathcal{F}(c_1^2, S) = 0,$$
$$\mathcal{F}(c_2^2, S) = 1,$$

and both of the target candidates get more vote than their opponents.

3.3. Objective Functions

In this paper, our goal is to maximize/minimize the number of winners from our target party. Then the objective functions are the same as [9]. Considering both IC and LT models, we define $\mathcal{F}(C_1, S)$ as the number of candidates in C_1 that are among the winners. Formally, consider a set of given activated nodes A_S, which became active after S. Let us define $\mathcal{F}_{A_S}(c)$ as the expected number of votes that candidate c will receive while A_S is the set of activated nodes. We set $\mathcal{Y}_{A_S}(c)$ as the number of candidates $c' \in C \setminus \{c\}$ where the expected number of their votes is less than c. In order to consider the tie-breaking rule, if $\mathcal{F}_{A_S}(c_i^j) = \mathcal{F}_{A_S}(c_{i'}^{j'})$, then c_i^j has more priority than $c_{i'}^{j'}$ if $j < j'$, or $j = j' \wedge i < i'$. Then $\mathcal{Y}_{A_S}(c)$ is defined as

$$\mathcal{Y}_{A_S}(c_i^j) = \left| \{c_{i'}^{j'} \in C \mid \mathcal{F}_{A_S}(c_i^j) > \mathcal{F}_{A_S}(c_{i'}^{j'}) \vee (\mathcal{F}_{A_S}(c_i^j) = \mathcal{F}_{A_S}(c_{i'}^{j'}) \wedge (j < j' \vee (j = j' \wedge i < i')))\} \right|.$$

By this definition, we define $\mathcal{F}(C_1, S)$ as the expected number of winners from party C_1, i.e., $\mathcal{F}(C_1, S) = \mathbb{E}_{A_S} \left[\sum_{c \in C_1} \mathbb{1}_{\mathcal{Y}_{A_S}(c) \geqslant (t-1)k} \right]$.

Now, let us define the first objective function as Difference of Winners (DoW), where is the difference between the number of winners in our target party before and after S. Formally, in constructive (resp., destructive) model we define DoW_c (resp., DoW_d) as

$$\text{DoW}_c(C_1, S) = \mathcal{F}(C_1, S) - \mathcal{F}(C_1, \emptyset),$$
$$\text{DoW}_d(C_1, S) = \mathcal{F}(C_1, \emptyset) - \mathcal{F}(C_1, S).$$

The problem of constructive difference of winners (CDW) asks for finding a set of seed nodes S ($|S| \leqslant B$) to maximize $\text{DoW}_c(C_1, S)$. Similarly, destructive difference of winners (DDW) refers to the problem of finding a set of seed node S ($|S| \leqslant B$) to maximize $\text{DoW}_d(C_1, S)$.

As the second objective function, we define a more compelling one called Margin of Victory (MoV). For constructive case, we define it as DoW plus the difference between the number of winners in the opponent parties with the most winners after and before S. Formally, for constructive (resp., destructive) case, we define MoV_c (resp., MoV_d) as

$$\text{MoV}_c(C_1, S) = \mathcal{F}(C_1, S) - \mathcal{F}(C_A^S, S) - (\mathcal{F}(C_1, \emptyset) - \mathcal{F}(C_B, \emptyset)),$$
$$\text{MoV}_d(C_1, S) = \mathcal{F}(C_1, \emptyset) - \mathcal{F}(C_B, \emptyset) - (\mathcal{F}(C_1, S) - \mathcal{F}(C_A^S, S)),$$

where C_B, C_A^S, respectively, are the opponent parties with the most winner before and after S.

The constructive margin of victory (CMV) problem is looking for a set of seed nodes S ($|S| \leqslant B$) in order to maximize $\text{MoV}_c(C_1, S)$. Similarly, destructive margin of victory (DMV) refers to the problem of finding a set of seed nodes S ($|S| \leqslant B$) to maximize $\text{MoV}_d(C_1, S)$.

4. Multi-Winner Election Control on Graph under LTM

It is proven that the problem is *NP*-hard to approximate within any factor of approximation using ICM [9]. In this part, we prove the same statement considering LTM.

Theorem 1. *It is NP-hard to approximate* CMV *and* CDW *within any factor on a given graph under LTM.*

Proof. Let us reduce the vertex cover (VC) problem to any approximation algorithm for CDW (reps., CMV). In VC, we are given an undirected graph $G = (V, E)$ and an integer k; the decision question is: Is there a set of nodes $V' \subseteq V$ ($|V'| \leqslant k$) so that for each edge $(u, v) \in E$, at least one of its vertices are in V'? Assume $\mathcal{I}(G, B)$ is a given instance for VC problem, where $G = (V, E)$ is the given graph, and B is an integer value. We create an instance $\mathcal{I}'(G', B)$ for CDW (reps., CMV) so that $G' = (V \cup V' \cup V'', E')$ is the graph build from G, and B is also the budget for our problem. Let us consider a case where there are two parties and four candidates, i.e., $t = k = 2, C = C_1 \cup C_2$, $C_1 = \{c_1^1, c_2^1\}, C_2 = \{c_1^2, c_2^2\}$. We fix the order of candidates in the probability distribution of the voter v as $\pi_v = (\pi_v(c_1^1), \pi_v(c_2^1), \pi_v(c_1^2), \pi_v(c_2^2))$, and build G' as follows.

- For each undirected edge $(u, v) \in E$ add two directed edges $(u, v), (v, u)$ to E'. Set the weight of each incoming edge to a node $v \in V$ as $\frac{1}{|N_v^i|}$. By this the sum over weight of all incoming edges is equal to one, i.e., $\forall v \in V : \sum_{u \in N_v^i} b_{u,v} = 1$.
- For each node $v \in V$, add two more nodes v', v'' to V', V'', respectively. Furthermore, add an edge (v, v') to E' with $b_{v,v'} = 1$. Formally, $\forall v \in V : v' \in V', v'' \in V'', (v, v') \in E'$ s.t. $b_{v,v'} = 1$. Note that nodes in V'' are isolated.
- Set the preferences list of the nodes as follows.

$$\forall v \in V, \pi_v = (\frac{1}{2}, \frac{1}{2}, 0, 0),$$

$$\forall v' \in V', \pi_{v'} = (\frac{1}{2}, 0, \frac{1}{2}, 0),$$

$$\forall v'' \in V'', \pi_{v''} = (0, 0, \frac{1}{2}, \frac{1}{2}).$$

By this reduction, the score of candidates before any diffusion is $\mathcal{F}(c_1^1, \emptyset) = \mathcal{F}(c_1^2, \emptyset) = |V|$, $\mathcal{F}(c_2^1, \emptyset) = \mathcal{F}(c_2^2, \emptyset) = \frac{1}{2}|V|$. Then $F(C_1, \emptyset) = \mathcal{F}(C_2, \emptyset) = 1$.

Note that in this reduction a node v will become active deterministically, if either it is selected as a seed node, or all of its incoming neighbors are selected as the seed nodes. Then if we can find a set of seed nodes $S \subseteq V$ so that it activates all nodes in V deterministically, the seed set S is also an answer for the corresponding VC problem.

In any approximation algorithm, we know that $S \subseteq V$ after the diffusion; otherwise, if there is a node $v' \in V' \cap S$ we can replace it with its incoming neighbor $v \in V$ such that $(v, v') \in E'$ and we get at least the same value for $\text{MoV}_c, \text{DoW}_c$. Furthermore, if there exists a node $v'' \in V'' \cap S$ one of the following situations holds:

- There exists an inactive node $v \in V \setminus A_S$ after the diffusion S. In this case, we can substitute v for v'' and then we get at least the same $\text{DoW}_c, \text{MoV}_c$.
- There is no inactive node $v \in V \setminus A_S$. In this case, according to the nodes' probability distribution, when all nodes in V become active, the value of MoV_c and DoW_c is maximum. Then even if we remove v'' from S it does not change the value of MoV_c or DoW_c. By the way, in this situation, if there exist any node $v \in V \setminus A_S$ we replace v'' with it, otherwise we replace it with a node $v \in V \setminus S$.

Then from now on, we assume $S \subseteq V$.

If all nodes in V become active, since they have an outgoing edge to all nodes $v' \in V'$ with probability one, then all nodes in $V \cup V'$ will become active, and the score of the candidates will be as follows.

$$\mathcal{F}(c_1^1, S) = |V|,$$

$$\mathcal{F}(c_2^1, S) = \mathcal{F}(c_1^2, S) = \frac{3}{4}|V|,$$

$$\mathcal{F}(c_2^2, S) = \frac{1}{2}|V|.$$

Then $F(C_1, S) = 2, \mathcal{F}(C_2, S) = 0$, $\text{DoW}_c(C_1, S) > 0$, $\text{MoV}_c(C_1, S) > 0$, and any approximation algorithm will return a positive value, then the answer of $\mathcal{I}$ will be YES.

On the other hand, if there is a node $v \in V$, which is inactive after the diffusion, i.e., $\exists v \in V \setminus A_S$, the score of candidates will be as follows.

$$\mathcal{F}(c_1^1, S) = |V|,$$

$$\mathcal{F}(c_2^1, S) < \frac{3}{4}|V|,$$

$$\mathcal{F}(c_1^2, S) > \frac{3}{4}|V|,$$

$$\mathcal{F}(c_2^2, S) = \frac{1}{2}|V|.$$

Then $F(C_1, S) = \mathcal{F}(C_2, S) = 1, \text{DoW}_c(C_1, S) = \text{MoV}_c(C_1, S) = 0$, and any approximation algorithm will return zero, then the answer of $\mathcal{I}$ will be NO.

For the other direction, note that if we can find a set of nodes $S \subseteq V$, which is an answer for $\mathcal{I}$, using the same set of nodes, we can activate all nodes in $V \cup V'$ and $\text{DoW}_c(C_1, S) > 0, \text{MoV}_c(C_1, S) > 0$.

To extend the proof for any number of parties (t) and candidates (k), we need to assign the probability distribution as follows, and the same approach concludes the proof for any $t, k > 2$. The same as before, the order of the candidates in probability distribution of a voter v is $\pi_v - (\pi_v(c_1^1), \ldots, \pi_v(c_k^1), \pi_v(c_1^2), \ldots, \pi_v(c_k^2), \quad , \pi_v(c_1^t), \ldots, \pi_v(c_k^t))$.

$$\forall v \in V, \pi_v = (\overbrace{\frac{1}{k}, \frac{1}{k}, \ldots, \frac{1}{k}}^{k}, \overbrace{0, \ldots, 0}^{k(t-1)}),$$

$$\forall v' \in V', \pi_{v'} = (\overbrace{\frac{1}{k}, \frac{1}{k}, \ldots, \frac{1}{k}}^{k-1}, 0, \frac{1}{k}, \overbrace{0, \ldots, 0}^{k(t-1)-1}),$$

$$\forall v'' \in V'', \pi_{v''} = (\overbrace{0, \ldots, 0}^{k}, \overbrace{\frac{1}{k}, \ldots, \frac{1}{k}}^{k}, \overbrace{0, \ldots, 0}^{k(t-2)}).$$

□

The following theorem proves the same statement for the destructive case of the problem.

Theorem 2. *It is NP-hard to approximate DMV and DDW within any factor on a given graph under LTM.*

Proof. The reduction is similar to the constructive case. Consider the case where $t = k = 2$. We should set the voters' probability distributions such that one of our target candidates be among the losers before and after any diffusion. Furthermore, another target candidate is among the winners before any dissemination; however, he will lose the election if and only if all nodes in the connected part of the graph become active. Please note that, since our target candidates have more priority than the others, we need one more node to be able to do that. □

5. Multi-Winner Election Control on Arborescence under ICM

In this section, instead of a general graph, we consider an arborescence structure. We are given a tree $G = (V, E)$ and a budget B where the directed edges are from leaves towards the root under ICM. We are asked to find at most B seed nodes to maximize MoV_c and DoW_c.

It has been shown that the problem in inapproximable on a general graph, except $P = NP$ [9]. Bharathi et al. conjectured that the IM problem considering ICM on arborescence is *NP*-hard [26]. Lu et al. proved that the conjecture is true [27], while Wang et al. showed that the IM problem accepts a polynomial-time algorithm on arborescence under LTM [28]. In the following, we show that our problem is hard to approximate within any factor of approximation on arborescence under ICM.

Theorem 3. *It is NP-hard to find an approximation algorithm for* CMV *and* CDW *on arborescence under ICM.*

Proof. We show the hardness by reducing the IM problem to our problem. Given an instance $\mathcal{I}(T, B)$ of IM problem where $T = (V, E)$ is the tree (arborescence), and B is the budget. Let us define the decision version of the problem as follows: is there at most B seed nodes so that it activates all nodes of the tree in expected?

We consider the case where there are two parties and each of them have just two candidates, i.e., $C = C_1 \cup C_2, C_1 = \{c_1^1, c_2^1\}, C_2 = \{c_1^2, c_2^2\}$. Furthermore, for simplicity, we consider the plurality scoring rule. The proof can be extended for any number of parties and candidates using any non-increasing scoring function, akin to [29].

Let us create an instance of our problem $\mathcal{I}'(T', B)$ as follows, where $T' = (V \cup V' \cup V'', E)$ is a tree, and B is the same budget for both problems.

- For each node $v \in V$ we add two more nodes v', v'' to V', V'', respectively, i.e., $\forall v \in V : v' \in V'$, $v'' \in V''$.
- For each node $v \in V$ we add an edge (v, v'') to E where $b_{v,v''} = 1$.
- Set the preference list of all nodes as follows.

$$\forall v \in V : c_1^2 \succ c_2^2 \succ c_1^1 \succ c_2^1,$$
$$\forall v' \in V' : c_2^2 \succ c_1^2 \succ c_2^1 \succ c_1^1,$$
$$\forall v'' \in V'' : c_1^2 \succ c_1^1 \succ c_2^1 \succ c_2^2$$

Clearly, seed nodes will be selected from V, i.e., $S \subseteq V$; otherwise, if there is a node $v' \in S \cap V'$, then the node is useless and does not affect DoW_c or MoV_c. If there is a node $v'' \in S \cap V''$, we can replace it with its incoming neighbor and get at least the same value for DoW_c and MoV_c.

Using aforementioned polynomial-time reduction, if there exists a set of nodes $S \subseteq V$ ($|S| \leqslant B$) so that $\text{MoV}_c > 0$ (resp. $\text{DoV}_c > 0$), then the node will activate all nodes in $V \cup V''$. Hence, we can select the same set and they will activate all nodes in T; then the answer of $\mathcal{I}$ will be YES. On the other hand, if $\text{MoV}_c = 0$ (resp. $\text{DoW}_c = 0$), it means there is no seed set can activate all nodes in $V \cup V''$; then the answer of $\mathcal{I}$ is NO. More formally, before any diffusion the score of candidates is

$$\mathcal{F}(c_1^1, \varnothing) = \mathcal{F}(c_2^1, \varnothing) = 0,$$
$$\mathcal{F}(c_1^2, \varnothing) = 2|V|,$$
$$\mathcal{F}(c_2^2, \varnothing) = |V|.$$

Then, none of the candidates in our target party will be elected as winner. After S, if there exists an inactive node in $V \cup V''$, then the the score of candidates will be as follows:

$$\mathcal{F}(c_1^1, S) < |V|,$$
$$\mathcal{F}(c_2^1, S) = 0,$$
$$\mathcal{F}(c_1^2, S) > |V|,$$
$$\mathcal{F}(c_2^2, S) = |V|.$$

In this case also, none of our target candidates will be among the winners, and $\text{MoV}_c = \text{DoW}_c = 0$. However, if all nodes in $V \cup V''$ become active after S, the score of the candidates will be as follows and one of our target candidates (c_1^1) will be elected as winner and any approximation algorithm will return $\text{MoV}_c > 0$ (resp. $\text{DoW}_c > 0$). It concludes the prove.

$$\mathcal{F}(c_1^1, S) = |V|,$$
$$\mathcal{F}(c_2^1, S) = 0,$$
$$\mathcal{F}(c_1^2, S) = |V|,$$
$$\mathcal{F}(c_2^2, S) = |V|.$$

□

The following theorem demonstrates the same hardness of approximation for the destructive case of our problem.

Theorem 4. *It is NP-hard to find an approximation algorithm for* DMV *and* DDW *on arborescence under ICM.*

Proof. The prove for the destructive case is similar to the constructive one. Consider $\mathcal{I}'$ in Theorem 3, we need to set the preferences list of the nodes so that all of our target candidates win the election before any diffusion; however, after the diffusion, one of them (let us say $c \in C_1$) will lose if and only if all nodes in $V \cup V''$ become active. Note that since our target candidates have more priority than the others, we need one more isolated node to ensure that c will lose the election after the diffusion. Following the same approach concludes the statement. □

6. Multi-Winner Election Control on Tree Using Straight-Party Voting

In this part, we consider the problem on a variation of the straight-party voting system (also called straight-ticket voting) in which the voters can vote for a party instead of candidates [30,31]. This model is used in many real elections [32,33]. The multi-winner election control problem via social influence under ICM and a general graph is considered in [9]. They showed that the problem is hard, and presented some constant factor approximation using straight-party voting system. In this section, we consider the problem on a tree where the edges are directed from root to the leaves.

In the rest of this section, we assume the given tree is a binary tree as we can convert any tree T to a binary tree T' by adding $O(n)$ fake nodes. However, our algorithm can use the fake nodes to navigate the tree, but they neither have a probability distribution (preference list) nor can be selected as a seed node. To ensure that the fake nodes will not change the diffusion process on the tree, the weight of each incoming edge to each fake node should be equal to one. Moreover, the weight of an edge from a fake node to an original node is equal to the weight of the original node's incoming edge in T.

In the following, we present some dynamic programming (DP) algorithm to maximize DoV_c^{spv} (and DoV_d^{spv}). Given a tree $T = (V, E)$, and budge B, the idea is that for a fixed node $v \in V$ and budget k $(0 \leqslant k \leqslant B)$, we calculate the maximum outcome from the sub-tree rooted at v, among the following cases: First, select the node v and try to find the other $k - 1$ seed nodes in its children. Second, do not select v and look for k seed nodes in its children.

We define $r(v), l(v), f(v)$, respectively, as the right child, left child, and the parent (father) of the node v. In Section 6.1 we consider the problem under LTM, and in Section 6.2 the problem is investigated under ICM.

6.1. Multi-Winner Election Control Using Straight-Party Voting under LTM

In this section, the voters have preferences list over the candidates. However, they vote for a party proportional to the probability of voting for all candidates in each party. Let us define $\mathcal{F}_{spv}(C_1, \varnothing), \mathcal{F}_{spv}(C_1, S)$, as the sum of the scores for our target party C_1 before and after S, respectively. Formally they are defined as follows.

$$\mathcal{F}_{spv}(C_1, \varnothing) = \mathbb{E}\Big[\sum_{v \in V} \sum_{c \in C_1} \pi_v(c) \Big],$$

$$\mathcal{F}_{spv}(C_1, S) = \mathbb{E}_{A_S}\Big[\sum_{v \in V} \sum_{c \in C_1} \tilde{\pi}_v(c) \Big].$$

The same as before we define the objective function MoV and difference of votes (DoV), for constructive case, as follows.

$$\mathrm{DoV}_c^{spv}(C_1, S) = \mathcal{F}_{spv}(C_1, S) - \mathcal{F}_{spv}(C_1, \varnothing),$$

$$\mathrm{MoV}_c^{spv}(C_1, S) = \mathcal{F}_{spv}(C_1, S) - \mathcal{F}_{spv}(C_A^S, S) - \big(\mathcal{F}_{spv}(C_1, \varnothing) - \mathcal{F}_{spv}(C_B, \varnothing)\big), \quad (1)$$

while C_B and C_A^S are the most voted opponent party before and after S, respectively. For destructive model the objective functions are defined as

$$\mathrm{DoV}_d^{spv}(C_1, S) = \mathcal{F}_{spv}(C_1, \varnothing) - \mathcal{F}_{spv}(C_1, S),$$

$$\mathrm{MoV}_d^{spv}(C_1, S) = \mathcal{F}_{spv}(C_1, \varnothing) - \mathcal{F}_{spv}(C_B, \varnothing) - \big(\mathcal{F}_{spv}(C_1, S) - \mathcal{F}_{spv}(C_A^S, S)\big). \quad (2)$$

6.1.1. Maximizing DoV in Straight-Party Voting under LTM

We define F_v as the set of possible probabilities that the node $f(v)$ may become active. More precisely, consider all nodes in the path from root to the v as $F_v' = \{v_0, v_1, \ldots, v_t = f(v)\}$ (recall that $f(v)$ is the parent of v). If none of the nodes in F_v' are selected as a seed node, then the probability that $f(v)$ becomes active by his incoming influence is zero. If just the root (v_0) is selected as the seed node, then the probability that $f(v)$ becomes active is $\prod_{i=0}^{i<t} b_{v_i, v_{i+1}}$; also, if v_1 is selected as a seed node but none of the nodes $v_i, 2 \leqslant i \leqslant t$, are selected as a seed node, the probability that $f(v)$ becomes active by its parent is $\prod_{i=1}^{i<t} b_{v_i, v_{i+1}}$, and so on; all these probabilities belong to F_v.

Let us define $\mathrm{DoV}_c(v, k, S, p)$ as the maximum value of the sum over the difference of probability to vote for our target party after and before S in the sub-tree rooted at v while $p \in F_v$ is the probability that its parent is active, and the budget is k. Furthermore, all selected seed nodes will be in S. In other words, $\mathrm{DoV}_c(v, k, S, p) = max\{\mathrm{DoV}_c^{spv}(C_1, S)\}$ in the sub-tree rooted at v while it will become active with probability $p \cdot b_{f(v), v}$ and $|S| \leqslant k$. The formal definition of $\mathrm{DoV}_c(v, k, S, p)$ is as follows:

$$\mathrm{DoV}_c(v, k, S, p) = max \Bigg\{$$

$$max_{k'=0}^{k} \Big\{ \mathrm{DoV}_c\big(r(v), k', S, p \cdot b_{f(v), v}\big) + \mathrm{DoV}_c\big(l(v), k - k', S, p \cdot b_{f(v), v}\big) \Big\} + p \cdot b_{f(v), v} \cdot \mathcal{D}_v,$$

$$max_{k'=0}^{k-1} \Big\{ \mathrm{DoV}_c\big(r(v), k', S \cup \{v\}, 1\big) + \mathrm{DoV}_c\big(l(v), k - k' - 1, S \cup \{v\}, 1\big) \Big\} + \mathcal{D}_v \Bigg\}, \quad (3)$$

where $\mathcal{D}_v$ is the increased score of our target party made by the node v if it becomes active, which is

$$\mathcal{D}_v = \sum_{c \in C_1} \left(\frac{\pi_v(c) + \frac{1}{|C_1|} \cdot p \cdot b_{f(v), v}}{1 + p \cdot b_{f(v), v}} - \pi_v(c) \right). \quad (4)$$

We can calculate and store the values in a two-dimensional array $A[B+1,|V|]$ where the rows are the budgets (starting from zero to B), and the columns are the nodes of the tree presented as the BFS reverse order, and each cell (i, j) $(0 \leqslant i \leqslant B, 0 \leqslant j < |V|)$ of the array refers to another array $A'[|F_{v_j}|]$. Then in the worst case, since the budget B, and $|F_{v_j}|$ (for any $v_j \in V$) are at most equal to $|V|$, then we can solve the problem in polynomial time using $O(|V|^3)$ memory. Note that we have to fill the matrix A left-to-right and top-down, while for each cell of it we can fill the corresponding array A' in any order.

As the base cases, for each leaf $v \in V$, and $p \in F_v$, if $k > 0$ we set $\mathrm{DoV}_c(v, k, S, p) = \mathcal{D}_v$, otherwise, if $k = 0$ we have $\mathrm{DoV}_c(v, k, S, p) = p \cdot b_{f(v),v} \cdot \mathcal{D}_v$ which is the difference of the probability to vote for our party after and before diffusion S, made by the node v. In fact, if the budget is greater than zero, the node will become active for sure, and we need to consider the difference of scores, but if the budget is zero we cannot select it as a seed node and the value should be multiplied by the probability that the node will become active, i.e., $p \cdot b_{f(v),v}$. We also define $\mathrm{DoV}_c(null, k, S, p) = 0$, that is, the value of DoV_c for a null reference is zero. It is useful when a node has just left (resp. right) child, then the value of the function for its right (resp. left) child, regardless of the other parameters, is zero. The pseudo-code of the DP is presented in Algorithm 1, which calculates the maximum DoV_c^{spv}; by small changes, it can find the seed nodes too. Note that the final answer will be calculated by $\mathrm{DoV}_c(v_{root}, B, \emptyset, 0)$ where v_{root} is the root node of the tree, B is the budget, $\emptyset$ represents that we have no seed node so far, and 0 means the parent of the root node will be activated with zero probability. The following theorem shows that the DP works well.

Algorithm 1: Calculating maximum DoV_c for e given tree T and budget B when the diffusion model is LTM and voting system is straight-party voting.

Procedure *DoV(Tree $T = (V, E)$, Budget B)*
    $A \leftarrow [B+1, |V|]$ ▷ It is a two-dimensional array $A[0..B, 0..|V|-1]$
    Name all nodes in V from 0 to $|V| - 1$ in BFS reverse order
    for *($j \leftarrow 0; j < |V|; j \leftarrow j + 1$)* **do**
       $F_{v_j} \leftarrow$ Set of all possible probabilities that $f(v_j)$ may become active
       for *($i \leftarrow 0; i <= B; i \leftarrow i + 1$)* **do**
            ▷ the variables i, j are a counter for rows and columns, respectively.
          $A[i, j] \leftarrow \mathrm{Array}[|F_{v_j}|]$ ▷ Each cell (i, j) is an array
          if *(v_j is a leaf)* **then**
             for *($p \in F_{v_j}$)* **do**
$$A[i, j; p] \leftarrow \sum_{c \in C_1} \left(\frac{\pi_{v_j}(c) + \frac{1}{|C_1|} \cdot p \cdot b_{f(v_j),v_j}}{1 + p \cdot b_{f(v_j),v_j}} - \pi_{v_j}(c) \right)$$
                if *($i = 0$)* **then**
                   $A[i, j; p] \leftarrow p \cdot b_{f(v_j),v_j} \cdot A[i, j; p]$
                end
             end
             continue
          end
          for *($p \in F_{v_j}$)* **do**
             ▷ If $r(v_j)$ or $l(v_j)$ does not exist, $A[..., r(v_j)$ or $l(v_j); ...]$ is zero.
$$\mathcal{D}_v \leftarrow \sum_{c \in C_1} \left(\frac{\pi_{v_j}(c) + \frac{1}{|C_1|} \cdot p \cdot b_{f(v_j),v_j}}{1 + p \cdot b_{f(v_j),v_j}} - \pi_{v_j}(c) \right)$$
$$max_j \leftarrow \max_{k=0}^i (A[k, r(v_j); p \cdot b_{f(v_j),v_j}] + A[i - k, l(v_j); p \cdot b_{f(v_j),v_j}])$$
$$max'_j \leftarrow \max_{k=0}^{i-1} (A[k, r(v_j); 1] + A[i - k - 1, l(v_j); 1])$$
$$A[i, j; p] \leftarrow \max(max_j + p \cdot b_{f(v_j),v_j} \cdot \mathcal{D}_v, max'_j + \mathcal{D}_v)$$
          end
       end
    end
    return $A[B, |V| - 1; 0]$ ▷ The final result for the root node using all budget
end

Theorem 5. *Given a tree $T = (V, E)$ and budget B, the DP Equation (3) finds a set of seed nodes S ($|S| \leqslant B$) to maximize DoV_c^{spv}.*

Proof. Consider the matrix $A[B + 1, |V|]$ where each cell $A[k, v]$ point to another array A' where the columns are all possible probabilities that $f(v)$ will become active. Calculating all possible probabilities for the array A', we have at most $|F_v|$ columns for each node $v \in V$ and budget $0 \leqslant k \leqslant B$, and for each of them, we need to calculate and store the maximum DoV_c.

Please note that if $f(v)$ becomes active, it can activate v with a probability equal to the weight of the edge between them ($b_{f(v),v}$). It holds because each node has just one incoming edge (its parent), and the threshold of the node will be generated uniformly at random. Then the probability that the threshold of the node v be less than (or equal) to the weight of the incoming edge is $b_{f(v),v}$.

Let us show that all values in the arrays will be calculated correctly, by induction. To see that, consider the base cases. For each leaf $v \in V$, the node cannot activate any other node as it has no outgoing edge. Then, these nodes cannot change the probability distribution of other nodes. In other words, each leaf will change just its own probability distribution. If $k = 0$, it means that we cannot select the node as a seed node, and we need to consider the probability of activating the node, because just activated nodes can update their probability distribution after the diffusion. Then if $k = 0$, we have $DoV_c(v, k, S, p) = p \cdot b_{f(v),v} \cdot \mathcal{D}_v$, where $\mathcal{D}_v$ is the difference of the party's score if the node v becomes active (defined in Equation (4)), and $p \cdot b_{f(v),v}$ is the probability that the node will be activated by its parent. On the other hand, if $k > 0$, we can select v as a seed node, and it will be activated with the probability of one, then we have $DoV_c(v, k, S, p) = \mathcal{D}_v$. Using the updating rule (defined in Section 3.1), and the definition of DoV_c^{spv} (defined in Equation (1)), the base cases are true.

Let us define $(i', j') < (i, j)$ if $j' < j$, or $j' = j \wedge i' < i$. We have shown that all arrays A' related to the base cases filled out correctly. Now by induction step, assume all related arrays related to pair (i', j') smaller than (i, j) are correctly calculated. In order to calculate the A' related to $A[i, j]$, for each column $p \in F_{v_j}$ we use following formula

$$
\begin{aligned}
DoV_c(v_j, i, S, p) = max \Big\{ \\
max_{k=0}^{i} \Big\{ DoV_c \left(r(v_j), k, S, p \cdot b_{f(v_j), v_j} \right) + DoV_c \left(l(v_j), i - k, S, p \cdot b_{f(v_j), v_j} \right) \Big\} + p \cdot b_{f(v_j), v_j} \cdot \mathcal{D}_{v_j}, \\
max_{k=0}^{i-1} \Big\{ DoV_c \left(r(v_j), k, S \cup \{v_j\}, 1 \right) + DoV_c \left(l(v_j), i - k - 1, S \cup \{v_j\}, 1 \right) \Big\} + \mathcal{D}_{v_j} \Big\},
\end{aligned}
$$

in which the first maximization considers the maximum value among all possible cases that we do not select the node v_j as a seed node, and the second one considers the maximum value among all possible cases that we choose v_j as a seed node. The last term in each maximization is the increased amount of DoV_c in the node v_j, which is according to the probability that v_j will become active. Note that in the above formula, we are using the value of DoV_c for the children of v_j, and the nodes are sorted as the BFS reverse order, then all required values are correctly calculated before, and we are selecting the maximum value among all possible cases. Then $DoV_c(v_j, i, S, p)$ will find the maximum possible value of DoV_c^{spv} correctly and concludes the proof. $\square$

For the destructive model, we define $DoV_d(v, k, S, p)$ as the maximum difference of probability to vote for our target party before and after S in the sub-tree rooted at v, while the budget is k and $p \in F_v$ is the probability that $f(v)$ will become active. Formally, we define $DoV_d(v, k, S, p)$ as follows.

$$\text{DoV}_d(v, k, S, p) = max \Bigg\{$$

$$max_{k'=0}^{k} \Big\{ \text{DoV}_d \left(r(v), k', S, p \cdot b_{f(v),v} \right) + \text{DoV}_d \left(l(v), k - k', S, p \cdot b_{f(v),v} \right) \Big\} + p \cdot b_{f(v),v} \cdot \mathcal{D}_v',$$

$$max_{k'=0}^{k-1} \Big\{ \text{DoV}_d \left(r(v), k', S \cup \{v\}, 1 \right) + \text{DoV}_d \left(l(v), k - k' - 1, S \cup \{v\}, 1 \right) \Big\} + \mathcal{D}_v' \Bigg\}, \quad (5)$$

where $\mathcal{D}_v' = \sum_{c \in C_1} \left(\pi_v(c) - \frac{\pi_v(c)}{1 + p \cdot b_{f(v),v}} \right)$ is the difference that the node v can apply. Moreover, for the base cases of the problem, for each leaf $v \in V$, and each probability $p \in F_v$, if $k = 0$ we need to consider the probability that the node will become active, then $\text{DoV}_d(v, k, S, p) = p \cdot b_{f(v),v} \cdot \mathcal{D}_v'$; otherwise, if $k > 0$, we have $\text{DoV}_d(v, k, S, p) = \mathcal{D}_v'$. Furthermore, we set $\text{DoV}_c(null, k, S, p) = 0$. The same as constructive case, for implementation we need a tow-dimensional array $A[B + 1, |V|]$. Moreover, for each cell $(i, j), 0 \leqslant i \leqslant B, 0 \leqslant j < |V|$, we keep another array $A'[|F_{v_j}|]$, where F_{v_j} is the set of possible probabilities that the node $f(v_j)$ can become active. The following theorem shows that by filling the matrix A left-to-right and up-down direction, we can find the optimal answer for DoV_d^{spv}.

Theorem 6. *Given a tree $T = (V, E)$ and a budget B, using the DP Equation (5), we can find a set of seed nodes S ($|S| \leqslant B$) to maximize DoV_d^{spv}.*

Proof. The proof is similar to Theorem 5, except for the base cases and the way of updating each activated node's probability distribution after the diffusion. Since a leaf cannot activate any other node, the only change that it can make is updating its own probability distribution. According to the updating rule (in Section 3.1), and the definition of DoV_d^{spv} (defined in Equation (2)), the base cases hold. Furthermore, by induction, we can see that the DP Equation (5) will find the maximum value of DoV_d^{spv} correctly. □

6.1.2. Maximizing MoV in Straight-Party Voting under LTM

In order to maximize MoV_c^{spv} we have to know C_A^S, i.e., the most voted opponent party after S. We have no problem to find the most voted opponent party before any diffusion (C_B); however, to find the most voted opponent party after S we need to have the optimal set of seed nodes that maximizes MoV_c^{spv}, and to find the optimal set of seed nodes we need the most voted opponent party (parties), which is a defective cycle.

To deal with this problem, someone may say that we consider $C_i, 2 \leqslant i \leqslant t$ as the most voted opponent party after S, and solve the related DP; after finding the outcome for all $t - 1$ parties, we select the maximum result as the output. Nevertheless, this is not true in all cases. Consider a case that there are two opponent parties, and each of them has half of the votes before any diffusion. If we consider each of them as the most voted opponent after the diffusion, we will get a wrong outcome as they both can be the most voted opponent after different diffusion processes. In fact, we need to consider multiple parties as the most voted opponent party.

By the way, it has been shown that by maximizing DoV_c^{spv} we get a $\frac{1}{3}$-approximation factor for maximizing MoV_c^{spv}. Moreover, by maximizing DoV_d^{spv} we get a $\frac{1}{2}$-approximation answer for maximizing MoV_d^{spv} [8].

6.2. Multi-Winner Election Control Using Straight-Party Voting under ICM

As we saw in previous section (in LTM), each node v becomes active either by being among the seed nodes or by the incoming influence from its parent $f(v)$. Since there is just one incoming edge for each node $v \in V$, and the threshold of the nodes t_v is generated uniformly at random, then the probability that its threshold be less than or equal to the incoming weight ($b_{f(v),v}$) is equal to $b_{f(v),v}$. In other words, the node will become active from its parent with the probability that its parent $f(v)$ is active, times the weight of the edge between them. On the other side, in ICM, a node v becomes active if it is either selected as a seed node or its parent $f(v)$ is activated and tries to influence v with the probability $b_{f(v),v}$. Then in a tree, the activation processes in both LTM and ICM are the same.

However, the updating rule is entirely different in them. In other words, in LTM, voters have a probability distribution over the candidates, and the activated nodes will update the probability of voting for candidates regarding the influence from activated incoming neighbors, while in ICM, voters have an exact preferences list over candidates, and the activated nodes promote/demote the position of some candidates in their preference list, regardless of neighbors (see Section 2 for a formal definition).

Since the diffusion process in ICM is the same as LTM, we focus more on updating part of the problem to maximize DoV_c^{spv}. Recall that we consider the plurality scoring rule for simplicity; however, it is possible to extend the results to any non-increasing scoring function. Then the scoring function $\mathcal{F}_{spv}$ for our target party is defined as follows. (To extend the result using any non-increasing scoring function $g(\cdot)$, we should define the functions as $\mathcal{F}_{spv}(C_1, \varnothing) = \sum_{v \in V} \sum_{c \in C_1} g(\pi_v(c))$, $\mathcal{F}_{spv}(C_1, S) = \mathbb{E}_{A_S}\left[\sum_{v \in V} \sum_{c \in C_1} g(\tilde{\pi}_v(c))\right]$.)

$$\mathcal{F}_{spv}(C_1, \varnothing) = \sum_{v \in V} \sum_{c \in C_1} \mathbb{1}_{\pi_v(c)=1},$$

$$\mathcal{F}_{spv}(C_1, S) = \mathbb{E}_{A_S}\left[\sum_{v \in V} \sum_{c \in C_1} \mathbb{1}_{\tilde{\pi}_v(c)=1}\right],$$

and the objective functions for the constructive and destructive cases of our problem are the same as Equations (1) and (2), respectively.

6.2.1. Maximizing DoV in Straight-Party Voting under ICM

In this case, node v can increase our target party's score by one, if none of our target candidates are in the first position before any diffusion, and one of them is in the second position of the voter's preference list. In other words, the voter v may increase the score of our target party if $\exists c \in C_1$, $\exists c' \in C \setminus C_1 : \pi_v(c') = 1 \wedge \pi_v(c) = 2$; otherwise, the node v can influence its children and change their opinion, but it cannot affect the target party's score. We call this condition as pre-condition and show it by $\P_v$. We define F_v as the set of all possible probabilities that the node v may become active (Please note that the definition of F_v in ICM is different from LTM). Consider a sub-tree rooted at $v \in V$, budget k, seed set S, and $p \in F_v$, we define $\mathrm{DoV}_c'(v, k, S, p)$ as follows.

$$\mathrm{DoV}_c'(v, k, S, p) = max\Big\{$$
$$max_{k'=0}^{k}\{\mathrm{DoV}_c'(r(v), k', S, p \cdot b_{v,r(v)}) + \mathrm{DoV}_c'(l(v), k - k', S, p \cdot b_{v,l(v)})\} + p \cdot \mathbb{1}_{\P_v},$$
$$max_{k'=0}^{k-1}\{\mathrm{DoV}_c'(r(v), k', S \cup \{v\}, b_{v,r(v)}) + \mathrm{DoV}_c'(l(v), k - k' - 1, S \cup \{v\}, b_{v,l(v)})\} + \mathbb{1}_{\P_v}\Big\}. \quad (6)$$

As the base cases of the problem, for each leaf $v \in V$, budget zero, and $p \in F_v$ as the probability that v will become active, we set $\mathrm{DoV}_c'(v, k, S, p) = p \cdot \mathbb{1}_{\P_v}$, and for the same parameters except a budget $k > 0$ we set $\mathrm{DoV}_c'(v, k, S, p) = \mathbb{1}_{\P_v}$. (To extend the algorithm for any non-increasing scoring function $g(\cdot)$, we need to define the base cases, respectively, as $\mathrm{DoV}_c'(v, k, S, p) = p \cdot (\sum_{c \in C_1, \exists c' \in C \setminus C_1 : \pi_v(c') < \pi_v(c)} g(\pi_v(c) - 1) - g(\pi_v(c)))$ and $\mathrm{DoV}_c'(v, k, S, p) = \sum_{c \in C_1, \exists c' \in C \setminus C_1 : \pi_v(c') < \pi_v(c)} g(\pi_v(c) - 1) - g(\pi_v(c))$.) The same as before, for each reference to a node which does not exists (*null*), we define $\mathrm{DoV}_c'(null, k, S, p) = 0$. In order to implement the DP Equation (6), the idea is the same as Algorithm 1. The following theorem shows that it calculates the maximum DoV_c^{spv} in polynomial-time.

Theorem 7. *Given a tree $T = (V, E)$, and budget B, the DP Equation (6) gives a set of seed nodes S ($|S| \leqslant B$) which maximizes DoV_c^{spv}.*

Proof. In DP Equation (6), there is a maximization over two other maximization formulae. The first one considers the case that we do not select v as a seed node; in this case, we consider the probability that node v will become active, i.e., $p \in F_v$. The second maximization considers selecting v as a seed node; in this state, v will be activated with probability equal to one. In both cases, the node may increase the function's value if the pre-condition holds; otherwise, it can influence its children. The same as previous proves, we show that it works by induction.

Consider a two-dimensional array $A[B + 1, |V|]$ where rows are the budgets from zero to B, and columns are the nodes in BFS reveres order. Each cell $A[i, j]$ ($0 \leqslant i \leqslant B, 0 \leqslant j < |V|$) refers to another array A' with the size of $|F_{v_j}|$. We calculate each array related to each cell (i, j) left-to-right and up-down direction.

To show that the base cases are correct, note that the leaves cannot activate any other node. Their only effect is by becoming active and changing their own opinion. Then there are two cases if the pre-condition holds for a leaf v: First, the budget is more than zero, then v can be a seed node and increase the amount of DoV'_c by one. Second, if the budget is zero, v can increment DoV'_c with the probability of becoming active through its parent, i.e., in expected, it will be $p \cdot \mathbb{1}_{\mathbb{q}_v}$ where $p \in F_v$ is the probability that v will be activated through its parent. Note that if the pre-condition does not hold, the leaf cannot make any effect, and in both cases, its effect is equal to zero.

Let us say $(i', j') < (i, j)$ if $j' < j$, or $j' = j \wedge i' < i$. As the step of induction, assume that all cells (i', j') smaller that (i, j) are filled correctly for $0 \leqslant i \leqslant B, 0 \leqslant j < |V|$. In order to calculate the array A' related to the cell (i, j), for each $p \in F_{v_j}$ we have to calculate the result of the following function.

$$
DoV'_c(v_j, i, S, p) = max \Big\{
$$
$$
max_{k=0}^{i} \{DoV'_c(r(v_j), k, S, p \cdot b_{v_j, r(v_j)}) + DoV'_c(l(v_j), i - k, S, p \cdot b_{v_j, l(v_j)})\} + p \cdot \mathbb{1}_{\mathbb{q}_v},
$$
$$
max_{k=0}^{i-1} \{DoV'_c(r(v_j), k, S \cup \{v_j\}, b_{v_j, r(v_j)}) + DoV'_c(l(v_j), i - k - 1, S \cup \{v_j\}, b_{v_j, l(v_j)})\} + \mathbb{1}_{\mathbb{q}_v} \Big\}.
$$

There is a maximization over two cases. Let us check each case separately. The first case considers all possible cases to split the budget into two parts for its children $r(v_j)$ and $l(v_j)$ (the first and second terms) when v_j is not selected as a seed node. It finds the split with the maximum outcome using the DoV'_c of its children, which are calculated correctly. In this case, since the node v_j is not a seed node, then the probability that its right (resp. left) child will become active is $p \cdot b_{v_j, r(v_j)}$ (resp. $p \cdot b_{v_j, l(v_j)}$). The fixed-term is the amount of change that the node v_j can afford to maximize our target party's score. If the pre-condition holds, then with the probability of p it will increase the score by one, that is $p \cdot \mathbb{1}_{\mathbb{q}_v}$.

The second maximization investigates the same situation except that it selects v_j as a seed node (if $i > 0$) and uses the value DoV'_c of its children to find the best split for the $i - 1$ remaining budgets. In this case, the node v_j can increase our party's score by one (if the pre-condition holds) as it is selected as a seed node and will be activated for sure. (To generalize the proof using any non-increasing scoring function $g(\cdot)$, we should change the updating part of each maximization (the fixed part) as $p \cdot (\sum_{c \in C_1, \exists c' \in C \setminus C_1 : \pi_v(c') < \pi_v(c)} g(\pi_v(c) - 1) - g(\pi_v(c)))$ and $\sum_{c \in C_1, \exists c' \in C \setminus C_1 : \pi_v(c') < \pi_v(c)} g(\pi_v(c) - 1) - g(\pi_v(c))$, respectively.) Note that all corresponding values for the children of v_j are correctly calculated before because the nodes are sorted as BFS reverse order. Finally, it finds the maximum value among the two cases. □

For the destructive case of the problem, we define pre-condition $\mathbb{q}'_v$ as $\exists c \in C_1 : \pi_v(c) = 1$. Then for a node v, if it becomes active and $\mathbb{q}'_v$ holds, the node will decrease the party's score by one; otherwise, v cannot change it. For each sub-tree rooted at v, budget k, and $p \in F_v$, let us define $DoV'_d(v, k, S, p)$ as follows.

$$\text{DoV}_d'(v,k,S,p) = max\Big\{$$
$$max_{k'=0}^{k}\{\text{DoV}_d'(r(v),k',S,p\cdot b_{v,r(v)}) + \text{DoV}_d'(l(v),k-k',S,p\cdot b_{v,l(v)})\} + p\cdot \mathbb{1}_{\P_v'},$$
$$max_{k'=0}^{k-1}\{\text{DoV}_d'(r(v),k',S\cup\{v\},b_{v,r(v)}) + \text{DoV}_d'(l(v),k-k'-1,S\cup\{v\},b_{v,l(v)})\} + \mathbb{1}_{\P_v'}\Big\}. \quad (7)$$

Note that the definition is exactly the same as constructive case except for the pre-condition. Furthermore the base cases are the same as before if we substitute $\P_v'$ for $\P_v$. The prove of the following theorem is similar to the Theorem 7; then we omit it to avoid repetition.

Theorem 8. *Given a tree $T = (V, E)$, and budget B, the DP Equation (7) gives a set of seed nodes S ($|S| \leqslant B$) which maximizes DoV_d^{spv}.*

6.2.2. Maximizing MoV in Straight-Party Voting under ICM

Similar to Section 6.1.2, we do not know the most scored parties after the diffusion started from a set of optimal seed nodes. However, it has been shown that by maximizing DoV_c^{spv} (resp. DoV_d^{spv}) we get a $\frac{1}{3}$ (resp. $\frac{1}{2}$) approximation algorithm for maximizing MoV_c^{spv} (resp. MoV_d^{spv}) [9].

7. Discussion

Controlling election via social influence is one of the most crucial parts of each democratic election. It has been shown that many campaigns are using this powerful tool to influence the voters and change their opinion during elections. In this work, we considered the multi-winner election control utilizing social influence so that the attacker tries to maximize/minimize the number of winners from his target party, concerning the party with the most winners.

We exhibited different results, including hardness of approximation, approximation guarantee, and optimal solutions for our problem considering different structures, diffusion models, and voting systems. In ICM, each voter has a preference list over the candidates and will vote for one or more candidate according to the voting rule, e.g., plurality, Borda's rule, k-approval, and anti-plurality. In this case, the influenced voters change their opinion by promoting/demoting the candidates' position in their preference list. On the other hand, in LTM, we consider that the voters have a probability distribution over all candidates. Each voter votes for one or more candidates proportional to the probability of voting for them. In this model, the activated voters change their opinion based on the incoming activated neighbors' influence.

We proved the problem is hard to approximate within any factor when the structure is a general graph, and the diffusion model is LTM. We also considered the problem when the structure is an arborescence, and the diffusion process follows the ICM rules. We showed that the problem is inapproximable within any factor, except $P = NP$. Another structure that we investigated is a tree where the voting system is a variation of straight-party voting. We presented a polynomial-time algorithm to maximize the expected score of our target party regarding both LT and IC diffusion models. It yields that we can get a $\frac{1}{3}$-approximation factor for maximizing MoV in constructive case, and $\frac{1}{2}$-approximation factor concerning MoV in the destructive model.

The results of this paper open several research directions. Considering the multi-winner election control through social influence on arborescence, when the diffusion model is LTM can be an exciting research problem. We conjecture that maximizing both objective functions (MoV and DoW) is hard; however, there exists a polynomial-time algorithm for the IM problem on arborescence under LTM. We plan to consider maximizing MoV in straight-party voting to either present an optimal solution or provide a hardness result regarding both constructive and destructive cases. Furthermore, maximizing DoV on the bidirected trees, where a child can activate its parent too, can be impressive.

We conjecture that the problem accepts a polynomial-time algorithm following a similar dynamic programming approach.

Author Contributions: Conceptualization, M.A.M. and G.D.; methodology, M.A.M. and G.D.; software, M.A.M. and G.D.; validation, M.A.M. and G.D.; formal analysis, M.A.M. and G.D.; investigation, M.A.M. and G.D.; resources, M.A.M. and G.D.; data curation, M.A.M. and G.D.; writing–original draft preparation, M.A.M. and G.D.; writing–review and editing, M.A.M. and G.D.; visualization, M.A.M and G.D.; supervision, G.D.; project administration, G.D.; funding acquisition, G.D. All authors have read and agreed to the published version of the manuscript.

Funding: This work has been partially supported by the Italian MIUR PRIN 2017 Project ALGADIMAR "Algorithms, Games, and Digital Markets".

Conflicts of Interest: The authors declare no conflict of interest.

References

1. Matsa, K.E.; Shearer, E. *News Use Across Social Media Platforms 2018*; Pew Research Center: Washington, DC, USA, 2018.
2. Bond, R.M.; Fariss, C.J.; Jones, J.J.; Kramer, A.D.I.; Marlow, C.; Settle, J.E.; Fowler, J.H. A 61-million-person experiment in social influence and political mobilization. *Nature* **2012**, *489*, 295.
3. Ferrara, E. Disinformation and social bot operations in the run up to the 2017 French presidential election. *First Monday* **2017**, *22*, doi:10.5210/fm.v22i8.8005.
4. Kreiss, D. Seizing the moment: The presidential campaigns' use of Twitter during the 2012 electoral cycle. *New Media Soc.* **2016**, *18*, 1473–1490.
5. Stier, S.; Bleier, A.; Lietz, H.; Strohmaier, M. Election Campaigning on Social Media: Politicians, Audiences, and the Mediation of Political Communication on Facebook and Twitter. *Political Commun.* **2018**, *35*, 50–74, doi:10.1080/10584609.2017.1334728.
6. Allcott, H.; Gentzkow, M. Social media and fake news in the 2016 election. *J. Econ. Perspect.* **2017**, *31*, 211–236.
7. Kempe, D.; Kleinberg, J.; Tardos, E. Maximizing the Spread of Influence through a Social Network. *Theory Comput.* **2015**, *11*, 105–147, doi:10.4086/toc.2015.v011a004.
8. Abouei Mehrizi, M.; Corò, F.; Cruciani, E.; D'Angelo, G. Election control through social influence with unknown preferences. In Proceedings of the 2020 International Computing and Combinatorics Conference, Atlanta, GA, USA, 29–31 August 2020; Springer: Berlin/Heidelberg, Germany, 2020; pp. 397–410.
9. Abouei Mehrizi, M.; D'Angelo, G. Multi-winner election control via social influence. In Proceedings of the Structural Information and Communication Complexity—27th International Colloquium (SIROCCO 2020), Paderborn, Germany, 29 June–1 July 2020; Richa, A.W., Scheideler, C., Eds.; Lecture Notes in Computer Science; Springer: Berlin/Heidelberg, Germany, 2020; Volume 12156, pp. 331–348, doi:10.1007/978-3-030-54921-3_19.
10. Wilder, B.; Vorobeychik, Y. Controlling elections through social influence. In Proceedings of the 17th International Conference on Autonomous Agents and MultiAgent Systems (AAMAS), Stockholm, Sweden, 10–15 July 2018; pp. 265–273.
11. Faliszewski, P.; Rothe, J.; Moulin, H. *Control and Bribery in Voting*; Handbook of Computational Social Choice; Cambridge University Press: Cambridge, UK, 2016; pp. 146–168.
12. Banerjee, S.; Jenamani, M.; Pratihar, D.K. A survey on influence maximization in a social network. *Knowl. Inf. Syst.* **2020**, *62*, 3417–3455.
13. Domingos, P.; Richardson, M. Mining the network value of customers. In Proceedings of the Seventh ACM SIGKDD International Conference on Knowledge Discovery and Data Mining, San Francisco, CA, USA, 26–29 August 2001; ACM: New York, NY, USA, 2001; pp. 57–66.
14. Richardson, M.; Domingos, P. Mining knowledge-sharing sites for viral marketing. In Proceedings of the 2002 ACM SIGKDD International Conference on Knowledge Discovery and Data Mining, Edmonton, AB, USA, 23–26 July 2001; ACM: New York, NY, USA, 2001; pp. 61–70.
15. Kempe, D.; Kleinberg, J.; Tardos, É. Maximizing the spread of influence through a social network. In Proceedings of the Ninth ACM SIGKDD International Conference on Knowledge Discovery and Data Mining, Washington, DC, USA, 24–27 August 2003; pp. 137–146.

16. Corò, F.; Cruciani, E.; D'Angelo, G.; Ponziani, S. Exploiting social influence to control elections based on scoring rules. In Proceedings of the 28th International Joint Conference on Artificial Intelligence (IJCAI), Macao, China, 10–16 August 2019.
17. Corò, F.; Cruciani, E.; D'Angelo, G.; Ponziani, S. Vote for me!: Election control via social influence in arbitrary scoring rule voting systems. In Proceedings of the 18th International Conference on Autonomous Agents and MultiAgent Systems (AAMAS '19), Montreal, QC, Canada, 13–17 May 2019; International Foundation for Autonomous Agents and Multiagent Systems: Richland, SC, USA, 2019; pp. 1895–1897.
18. Bredereck, R.; Elkind, E. Manipulating opinion diffusion in social networks. In Proceedings of the 26th International Joint Conference on Artificial Intelligence (IJCAI), Melbourne, Australia, 19–25 August 2017; pp. 894–900.
19. Castiglioni, M.; Ferraioli, D.; Gatti, N. Election control in social networks via edge addition or removal. In Proceedings of the 2020 AAAI Conference on Artificial Intelligence, New York, NY, USA, 7–12 February 2020; pp. 1878–1885.
20. Faliszewski, P.; Gonen, R.; Koutecký, M.; Talmon, N. Opinion diffusion and campaigning on society graphs. In Proceedings of the 27th International Joint Conference on Artificial Intelligence (IJCAI), Stockholm, Sweden, 13–19 July 2018; pp. 219–225.
21. Auletta, V.; Caragiannis, I.; Ferraioli, D.; Galdi, C.; Persiano, G. Minority becomes majority in social networks. In Proceedings of the 11th Web and Internet Economics (WINE), Amsterdam, The Netherlands, 9–12 December 2015; Springer: Berlin/Heidelberg, Germany, 2015; pp. 74–88.
22. Brill, M.; Elkind, E.; Endriss, U.; Grandi, U. Pairwise diffusion of preference rankings in social networks. In Proceedings of the 25th International Joint Conference on Artificial Intelligence (IJCAI), New York, NY, USA, 9–15 July 2016; pp. 130–136.
23. Botan, S.; Grandi, U.; Perrussel, L. Propositionwise opinion diffusion with constraints. In Proceedings of the 4th AAMAS Workshop (EXPLORE), Sao Paulo, Brazil, 8 May 2017.
24. Li, Y.; Fan, J.; Wang, Y.; Tan, K. Influence Maximization on Social Graphs: A Survey. *IEEE Trans. Knowl. Data Eng.* **2018**, *30*, 1852–1872, doi:10.1109/TKDE.2018.2807843.
25. Brandt, F.; Conitzer, V.; Endriss, U.; Lang, J.; Procaccia, A.D. *Handbook of Computational Social Choice*, 1st ed.; Cambridge University Press: New York, NY, USA, 2016.
26. Bharathi, S.; Kempe, D.; Salek, M. Competitive influence maximization in social networks. In *International Workshop on Web and Internet Economics*; Springer: Berlin/Heidelberg, Germany, 2007; pp. 306–311.
27. Lu, Z.; Zhang, Z.; Wu, W. Solution of Bharathi–Kempe–Salek conjecture for influence maximization on arborescence. *J. Comb. Optim.* **2017**, *33*, 803–808.
28. Wang, A.; Wu, W.; Cui, L. On Bharathi–Kempe–Salek conjecture for influence maximization on arborescence. *J. Comb. Optim.* **2016**, *31*, 1678–1684.
29. Abouei Mehrizi, M.; D'Angelo, G. Multi-Winner Election Control via Social Influence. *arXiv* **2020**, arXiv:2005.04037.
30. Campbell, B.A.; Byrne, M.D. Straight-party voting: What do voters think? *IEEE Trans. Inf. Forensics Secur.* **2009**, *4*, 718–728.
31. Kritzer, H.M. Roll-Off in State Court Elections: Change Over Time and the Impact of the Straight-Ticket Voting Option. *J. Law Court* **2016**, *4*, 409–435.
32. Ruhl, J.; Mcdonald, R. *Party Politics And Elections In Latin America*; Taylor & Francis: Abingdon, UK, 2019.
33. Hershey, M. *Party Politics in America*; Taylor & Francis: Abingdon, UK, 2017.

Article

On Multidimensional Congestion Games [†]

Vittorio Bilò [1],*, Michele Flammini [2],*, Vasco Gallotti [3] and Cosimo Vinci [2],*

1   Department of Mathematics and Physics "Ennio De Giorgi", University of Salento-Provinciale
    Lecce-Arnesano, P.O. Box 193, 73100 Lecce, Italy
2   Gran Sasso Science Institute-Viale Francesco Crispi 7, 67100 L'Aquila, Italy
3   Department of Information Engineering Computer Science and Mathematics, University of L'Aquila-Via
    Vetoio, Loc. Coppito, 67100 L'Aquila, Italy; vasco.gallotti@univaq.it
*   Correspondence: vittorio.bilo@unisalento.it (V.B.); michele.flammini@gssi.it (M.F.);
    cosimo.vinci@gssi.it (C.V.)
†   This work widely improves on an extended abstract presented at SIROCCO 2012.

Received: 9 September 2020; Accepted: 3 October 2020; Published: 15 October 2020

Abstract: We introduce multidimensional congestion games, that is, congestion games whose set of players is partitioned into $d + 1$ clusters $C_0, C_1, \ldots, C_d$. Players in C_0 have full information about all the other participants in the game, while players in C_i, for any $1 \leq i \leq d$, have full information only about the members of $C_0 \cup C_i$ and are unaware of all the others. This model has at least two interesting applications: (i) it is a special case of graphical congestion games induced by an undirected social knowledge graph with independence number equal to d, and (ii) it represents scenarios in which players have a type and the level of competition they experience on a resource depends on their type and on the types of the other players using it. We focus on the case in which the cost function associated with each resource is affine and bound the price of anarchy and stability as a function of d with respect to two meaningful social cost functions and for both weighted and unweighted players. We also provide refined bounds for the special case of $d = 2$ in presence of unweighted players.

Keywords: congestion games; pure Nash equilibrium; potential games; price of anarchy; price of stability

---

1. Introduction

Congestion games [1–4] are, perhaps, the most famous class of non-cooperative games due to their capability to model several interesting competitive scenarios, while maintaining nice properties. In these games, there is a set of players sharing a set of resources. Each resource has an associate cost function which depends on the number of players using it (the so-called *congestion*). Players aim at choosing subsets of resources so as to minimize the sum of the resource costs.

Congestion games were introduced by Rosenthal in Reference [2]. He proved that each such a game admits a bounded potential function whose set of local minima coincides with the set of *pure Nash equilibria* [5] of the game, that is, strategy profiles in which no player can decrease her cost by unilaterally changing her strategic choice. This existence result makes congestion games particularly appealing especially in all those applications in which pure Nash equilibria are elected as the ideal solution concept.

In these contexts, the study of inefficiency due to selfish and non-cooperative behavior has affirmed as a fervent research direction. To this aim, the notions of *price of anarchy* [6] and *price of stability* [7] are widely adopted. The price of anarchy (resp. stability) compares the performance of the worst (resp. best) pure Nash equilibrium with that of an optimal cooperative solution.

Congestion games with unrestricted cost functions are general enough to model the Prisoner's Dilemma game, whose unique pure Nash equilibrium is known to perform arbitrarily bad with respect to the solution in which all players cooperate. Hence, in order to deal with significative bounds on the prices of anarchy and stability, some kind of regularity needs to be imposed on the cost functions associated with the resources. To this aim, lot of research attention has been devoted to the case of polynomial cost functions [8–17], and more general latency functions verifying some mild assumptions [10,18,19]. Among these, the particular case of affine functions occupies a predominant role. In fact, as shown in References [20–22], they represent the only case, together with that (perhaps not particularly meaningful) of exponential cost functions, for which *weighted congestion games* [20], that is the generalization of congestion games in which each player has a weight and the congestion of a resource becomes the sum of the weights of its users, still admit a potential function.

1.1. Motivations

Traditional (weighted) congestion games are defined under a *full information* scenario—each player knows all the other participants in the game as well as their available strategies. These requirements, anyway, become too strong in many practical applications, where players may be unaware about even the mere existence of other potential competitors. This observation, together with the widespread of competitive applications in social networks, has drawn some attention on the model of *graphical (weighted) congestion games* [23–25].

A graphical (weighted) congestion game $(\mathcal{G}, G)$ is obtained by coupling a traditional (weighted) congestion game $\mathcal{G}$ with a *social knowledge graph* G expressing the *social context* in which the players operate. In G, each node is associated with a player of $\mathcal{G}$ and there exists a directed edge from node i to node j if and only if player i has full information about player j. A basic property of (weighted) congestion games is that the congestion of a resource r in a given strategy profile σ is the same for all players. The existence of a social context in graphical (weighted) congestion games, instead, makes the congestion of each resource player dependent. In these games, in fact, the congestion presumed by player i on resource r in the strategy profile σ is obtained by excluding from the set of players choosing r in σ those of whom player i has no knowledge. Clearly, if G is complete, then there is no difference between $(\mathcal{G}, G)$ and $\mathcal{G}$. In all the other cases, however, there may be a big difference between the cost that a player *presumes* to pay on a certain strategy profile and the real cost that she effectively *perceives* because of the presence of players she was unaware of during her decisional process (We observe that the model of graphical congestion games is sufficiently powerful to describe an alternative scenario in which players never perceive their real costs, which are perceived and evaluated by a central entity only. In such case, the central entity aims at evaluating the global and real impact on the performance of the game caused by the players' strategic behaviour).

Graphical congestion games have been introduced by Bilò et al. in Reference [24]. They focus on affine cost functions and provide a complete characterization of the cases in which existence of pure Nash equilibria can be guaranteed. In particular, they show that equilibria always exist if and only if G is either undirected or directed acyclic. Then, for all these cases, they give bounds on the price of anarchy and stability expressed as a function of the number of players in the game and of the maximum degree of G. These metrics are defined with respect to both the sum of the perceived costs and the sum of the presumed ones.

Fotakis et al. [25] argue that the maximum degree of G is not a proper measure of the level of social ignorance in a graphical congestion game and propose to bound the prices of anarchy and stability as a function of the independence number of G, denoted by $\delta(G)$. They focus on graphical weighted congestion games with affine cost functions and show that they still admit a potential function when G is undirected. Then, they prove that the price of anarchy is between $\delta(G)(\delta(G)+1)$ and $\delta(G)(\delta(G)+2+\sqrt{\delta(G)^2+4\delta(G)})/2$ with respect to both the sum of the perceived costs and the sum of the presumed ones, and that the price of stability is between $\delta(G)$ and $2\delta(G)$ with respect to the sum of the perceived costs.

1.2. Our Contribution and Significance

The works of Bilò et al. [24] and Fotakis et al. [25] aim at characterizing the impact of social ignorance in the most general case, that is, without imposing any particular structure on the social knowledge graphs defining the graphical game. Nevertheless, real-world-based knowledge relationships usually obey some regularities and present recurrent patterns: for instance, people tend to cluster themselves into well-structured collaborative groups (cliques) due to family memberships, mutual friendships, interest sharing, business partnerships.

To this aim, we introduce and study *multidimensional (weighted) congestion games*, that is, (weighted) congestion games whose set of players are partitioned into $d + 1$ clusters $C_0, C_1, \ldots, C_d$. Players in C_0 have full information about all the other participants in the game, while players in C_i, for any $1 \leq i \leq d$, have full information only about the members of $C_0 \cup C_i$ and are unaware of all the others. It is not difficult to see (and we provide a formal proof of this fact in the next section) that each multidimensional (weighted) congestion game is a graphical (weighted) congestion game whose social knowledge graph G is undirected and verifies $\delta(G) = d$. In addition, G possesses the following, well-structured, topology: it is the union of $d + 1$ disjoint cliques (each corresponding to one of the $d + 1$ clusters in the multidimensional (weighted) congestion game) with the additional property that there exists an edge from all the nodes belonging to one of these cliques (the one corresponding to cluster C_0) to all the nodes in all the other cliques.

We believe that the study of graphical games restricted to some prescribed social knowledge graphs like the ones we consider here, may be better suited to understand the impact of social ignorance in non-cooperative systems coming from practical and real-world applications. Moreover, the particular social knowledge relationships embedded in the definition of multidimensional (weighted) congestion games perfectly model the situation that generates when several independent games with full information are gathered together by some promoting parties so as to form a sort of "global super-game". The promoting parties become players with full information in the super-game, while each player in the composing sub-games maintains full information about all the other players in the same sub-game, acquires full information about all the promoting parties in the super game, but completely ignores the players in the other sub-games. Such a composing scheme resembles, in a sense, the general architecture of the Internet, viewed as a self-emerged network resulting from the aggregation of several autonomous systems (AS). Users in an AS have full information about anything happening within the AS, but, at the same time, they completely ignore the network global architecture and how it develops outside their own AS, except for the existence of high-level network routers. High-level network routers, instead, have full information about the entire network.

Furthermore, multidimensional (weighted) congestion games are also useful to model games in which players belong to different types and the level of competition that each player experiences on each selected resource depends on her type and on the types of the other players using the resource. Consider, for instance, a machine which is able to perform d different types of activities in parallel and a set of tasks requiring the use of this machine. Tasks are of two types: simple and complex. Simple tasks take the machine busy on one particular activity only, while complex tasks require the completion of all the d activities. Hence, complex tasks compete with all the other tasks, while simple ones compete only with the tasks requiring the same machine (thus, also with complex tasks). A similar example is represented by a facility location game where players want to locate their facilities so as to minimize the effect of the competition due to the presence of neighbor competitors. If we assume that the facilities can be either specialized shops selling particular products (such as perfumeries, clothes shops, shoe shops) or shopping centers selling all kinds of products, we have again that the shopping centers compete with all the other participants in the game, while specialized shops compete only with shops of the same type and with shopping centers.

In this paper, we focus on multidimensional (weighted) congestion games with affine cost functions. In such a setting, we bound the price of anarchy and the price of stability with respect to the two social cost functions, which are the sum of the perceived costs and the sum of the presumed costs. In fact, when multidimensional (weighted) congestion games are viewed as graphical (weighted) congestion games with highly clustered knowledge relationships, the sum of the perceived costs is more appropriate to define the overall quality of a profile: players decide according to their knowledge, but then, when the solution is physically realized, their cost becomes influenced also by the players of which they were not aware. Hence, under this social cost function, the notions of price of anarchy and price of stability effectively measure the impact of social ignorance in the system. On the other hand, when multidimensional (weighted) congestion games are used to model players belonging to different types, the perceived cost of a player coincides with the presumed one since there is no real social ignorance, even if the fact that players can be of different types allows for a reinterpretation of the model as a special case of graphical (weighted) congestion games. Hence, in such a setting, the inefficiency due to selfish behavior has to be analyzed with respect to the sum of the presumed costs.

We determine general upper bounds for the price of anarchy and the price of stability as a function of d. For the sum of the presumed costs, we show that the price of anarchy and stability of weighted games are at most $\frac{(\sqrt{d+4}+\sqrt{d})(\sqrt{d}\sqrt{d+4}+d+4)}{4\sqrt{d+4}} \leq d+2$ and 2, respectively. Instead, for the sum of the perceived costs, the results of Reference [25] yield upper bounds of $d(d+2+\sqrt{d^2+4d})/2$ and $2d$ for the price of anarchy and the price of stability, respectively.

Then, we investigate the case of unweighted games with $d=2$ (i.e., bidimensional congestion games) in higher depth and provide refined bounds. In particular, we prove that price of anarchy is $119/33 \approx 3.606$ with respect to the sum of the presumed costs and it is $35/8 = 4.375$ with respect to the sum of the perceived ones, and that the price of stability is between $(1+\sqrt{5})/2 \approx 1.618$ and $1+2/\sqrt{7} \approx 1.756$ for the sum of the presumed costs as social cost function, and between $(5+\sqrt{17})/4 \approx 2.28$ and 2.92 for the sum of the perceived ones. These results are derived by exploiting the primal-dual method developed in Reference [11].

A preliminary version of this paper has been presented at SIROCCO 2012 [26].

1.3. Paper Organization

Next section contains all formal definitions and notation. In Section 3 we discuss the existence of pure Nash equilibria in multidimensional weighted congestion games. In Sections 4 and 5, we present our bounds for the price of anarchy and the price of stability, respectively. Finally, in the last section, we give some concluding remarks and discuss open problems.

2. Model and Definitions

For an integer $n \geq 2$, we denote $[n] := \{1,2,\ldots,n\}$. In a *weighted congestion game* $\mathcal{G}$, we have n players and a set of resources R, where each resource $r \in R$ has an associated cost function ℓ_r. The set of strategies for each player $i \in [n]$, denoted as S_i, can be any subset of the powerset of R, that is, $S_i \subseteq 2^R$. Each player $i \in [n]$ is associated with a positive weight $w_i > 0$. Given a strategy profile $\sigma = (\sigma_1,\ldots,\sigma_n)$, the congestion of resource r in σ, denoted as $n_r(\sigma)$, is the total weight of players choosing r in σ, that is, $n_r(\sigma) = \sum_{i \in [n]:r \in \sigma_i} w_i$. The perceived cost paid by player i in σ is $c_i(\sigma) = \sum_{r \in \sigma_i} \ell_r(n_r(\sigma))$. An *unweighted congestion game* (congestion game, for brevity) is a weighted congestion game in which all players have unitary weights. An *affine weighted congestion game* is a weighted congestion game such that, for any $r \in R$, it holds that $\ell_r(x) = \alpha_r x + \beta_r$, with $\alpha_r, \beta_r \geq 0$; the game is *linear* if $\beta_r = 0$ for any $r \in R$.

For any integer $d \geq 2$, a *d-dimensional weighted congestion game* $(\mathcal{G},\mathcal{C})$ consists of a weighted congestion game $\mathcal{G}$ whose set of players is partitioned into $d+1$ clusters $C_0, C_1, \ldots, C_d$. For a player i, we denote by $f(i) \in \{0,\ldots,d\}$ the cluster she belongs to. We say that players in C_0 are *omniscient* and that players in C_i, for any $i \in [d]$, are *ignorant*. Given a strategy profile σ, we denote by $n_{r,j}(\sigma)$ the total weight of players belonging to C_j who are using resource r in σ. The presumed cost of a player i in σ is $\hat{c}_i(\sigma) = \sum_{r \in \sigma_i} \ell_r(n_{r,f(i)}(\sigma) + n_{r,0}(\sigma))$ if she is ignorant and $\hat{c}_i(\sigma) = \sum_{r \in \sigma_i} \ell_r(\sum_{j=0}^{d} n_{r,j}(\sigma)) =$

$\sum_{r \in \sigma_i} \ell_r(n_r(\sigma)) = c_i(\sigma)$ if she is omniscient. A *d-dimensional weighted affine congestion game* is a pair $(\mathcal{G}, \mathcal{C})$ such that $\mathcal{G}$ is an affine weighted congestion game.

Given a strategy profile σ and a strategy $s \in S_i$ for a player $i \in [n]$, we denote with (σ_{-i}, s) the strategy profile obtained from σ by replacing the strategy σ_i played by i in σ with s. A *pure Nash equilibrium* is a strategy profile σ such that, for any player $i \in [n]$ and for any strategy $s \in S_i$, it holds that $\hat{c}_i(\sigma_{-i}, s) \geq \hat{c}_i(\sigma)$.

Let Σ be the set of all possible strategy profiles which can be realized in $(\mathcal{G}, \mathcal{C})$. We denote with $\mathcal{NE}(\mathcal{G}, \mathcal{C}) \subseteq \Sigma$ the set of pure Nash equilibria of $(\mathcal{G}, \mathcal{C})$. Let $\mathrm{SF} : \Sigma \to \mathbb{R}_{\geq 0}$ be a *social cost function* measuring the overall quality of each strategy profile in Σ. We denote with σ^* a *social optimum* of $(\mathcal{G}, \mathcal{C})$ with respect to SF, that is, a strategy profile minimizing the social cost function SF. We consider two social cost functions, namely, the (weighted) sum of the presumed costs of all players and the (weighted) sum of their perceived ones denoted, respectively, as Pres and Perc. Technically, they assume the following expressions:

$$
\begin{aligned}
\mathrm{Pres}(\sigma) &= \sum_{i \in [n]} w_i \hat{c}_i(\sigma) \\
&= \sum_{i \in C_0} w_i \sum_{r \in \sigma_i} (\alpha_r n_r(\sigma) + \beta_r) + \sum_{i \notin C_0} w_i \sum_{r \in \sigma_i} \left(\alpha_r \left(n_{r,f(i)}(\sigma) + n_{r,0}(\sigma) \right) + \beta_r \right) \\
&= \sum_{r \in R} \left(\alpha_r n_{r,0}(\sigma) \sum_{j=0}^{d} n_{r,j}(\sigma) + \beta_r n_{r,0}(\sigma) \right) \\
&\quad + \sum_{r \in R} \left(\alpha_r n_{r,0}(\sigma) \sum_{j=1}^{d} n_{r,j}(\sigma) + \alpha_r \sum_{j=1}^{d} n_{r,j}(\sigma)^2 + \beta_r \sum_{j=1}^{d} n_{r,j}(\sigma) \right) \\
&= \sum_{r \in R} \left(\alpha_r \left(\sum_{j=0}^{d} n_{r,j}(\sigma)^2 + 2 n_{r,0}(\sigma) \sum_{j=1}^{d} n_{r,j}(\sigma) \right) + \beta_r \sum_{j=0}^{d} n_{r,j}(\sigma) \right)
\end{aligned}
$$

and

$$
\mathrm{Perc}(\sigma) = \sum_{i \in [n]} w_i c_i(\sigma) = \sum_{i \in [n]} w_i \sum_{r \in \sigma_i} (\alpha_r n_r(\sigma) + \beta_r) = \sum_{r \in R} \left(\alpha_r n_r(\sigma)^2 + \beta_r n_r(\sigma) \right).
$$

For a fixed social cost function SF, the *price of anarchy* of $(\mathcal{G}, \mathcal{C})$, denoted by $PoA(\mathcal{G}, \mathcal{C})$, is the ratio between the social value of the *worst* pure Nash equilibrium of $(\mathcal{G}, \mathcal{C})$ and that of a social optimum, that is, $PoA(\mathcal{G}, \mathcal{C}) = \max_{\sigma \in \mathcal{NE}(\mathcal{G}, \mathcal{C})} \frac{\mathrm{SF}(\sigma)}{\mathrm{SF}(\sigma^*)}$. The *price of stability*, denoted by $PoS(\mathcal{G}, \mathcal{C})$, considers the best case, so that $PoS(\mathcal{G}, \mathcal{C}) = \min_{\sigma \in \mathcal{NE}(\mathcal{G}, \mathcal{C})} \frac{\mathrm{SF}(\sigma)}{\mathrm{SF}(\sigma^*)}$.

3. Existence of Pure Nash Equilibria

In this section, we prove that multidimensional unweighted (resp. weighted affine) congestion games are graphical unweighted (resp. weighted affine) congestion games defined by an underlying undirected social knowledge graph. This allows us to use a result in Reference [24] (resp. [25]) stating that these games are potential games, thus admitting pure Nash equilibria.

A *graphical weighted congestion game* $(\mathcal{G}, G)$ consists of a weighted congestion game $\mathcal{G}$ and a directed graph $G = (N, A)$ such that each node of N is associated with a player in $\mathcal{G}$ and there exists a directed edge from node i to node j if and only if player i has full information about player j. The congestion presumed by player i on resource r in the profile σ is $\tilde{n}_{r,i}(\sigma) = \sum_{p \in N : r \in \sigma_p, (i,p) \in A} w_p + w_i$ and the presumed cost paid by player i in σ is $\tilde{c}_i(\sigma) = \sum_{r \in \sigma_i} \ell_r(\tilde{n}_{r,i}(\sigma))$. A *graphical weighted affine congestion game* is a pair $(\mathcal{G}, G)$ such that $\mathcal{G}$ is an affine weighted congestion game. The *independence number* $\delta(G)$ of $(\mathcal{G}, G)$ is the cardinality of a maximum independent set of graph G.

A function $\Phi : \Sigma \to \mathbb{R}$ is a *weighted potential function* for a graphical weighted congestion game $(\mathcal{G}, G)$, if for any profile σ, any player $i \in [n]$ and any strategy $s \in S_i$, it holds that $\Phi(\sigma) - \Phi(\sigma_{-i}, s) = a_i(\tilde{c}_i(\sigma) - \tilde{c}_i(\sigma_{-i}, s))$ for some $a_i > 0$; if $a_i = 1$, Φ is an *exact potential function*. In Reference [24] (resp. [25]), it is shown that each graphical unweighted (resp. weighted affine) congestion game $(\mathcal{G}, G)$ such that G is undirected admits an exact potential function (resp. weighted potential function).

The following result shows that d-dimensional weighted congestion games are a subclass of graphical weighted congestion games.

Proposition 1. *Each d-dimensional weighted congestion game is a graphical weighted congestion game whose social knowledge graph is undirected.*

Proof. Fix a d-dimensional weighted congestion game $(\mathcal{G}, \mathcal{C})$. We define a graph $G = (N, A)$ such that each node in N is associated with a player in $\mathcal{G}$ and there is an undirected edge $\{u, v\} \in A$ if and only if either $u, v \in C_i$ for some $0 \leq i \leq d$ or $u \in C_0$. We show that, for any strategy profile σ of $\mathcal{G}$ and for any $i \in [n]$, $\hat{c}_i(\sigma) = \tilde{c}_i(\sigma)$.

Consider first an omniscient player $i \in C_0$. In $(\mathcal{G}, \mathcal{C})$, it holds that

$$\hat{c}_i(\sigma) = \sum_{r \in \sigma_i} \ell_r(n_r(\sigma)) = \sum_{r \in \sigma_i} \ell_r \left(\sum_{p \in [n]: r \in \sigma_p} w_p \right),$$

while in $(\mathcal{G}, G)$, it holds that

$$\tilde{c}_i(\sigma) = \sum_{r \in \sigma_i} \ell_r(\tilde{n}_{r,i}(\sigma)) = \sum_{r \in \sigma_i} \ell_r \left(\sum_{p \in N: r \in \sigma_p, \{i,p\} \in A} w_p + w_i \right) = \sum_{r \in \sigma_i} \ell_r \left(\sum_{p \in [n]: r \in \sigma_p} w_p \right),$$

where the last equality follows from the fact that, by construction of G, it holds that $\{i, p\} \in A$, for any $p \in [n]$ with $p \neq i$.

Next, consider an ignorant player $i \in C_j$ for some $j \in [d]$. In $(\mathcal{G}, \mathcal{C})$, it holds that

$$\hat{c}_i(\sigma) = \sum_{r \in \sigma_i} \ell_r(n_{r,f(i)}(\sigma) + n_{r,0}(\sigma)) = \sum_{r \in \sigma_i} \ell_r \left(\sum_{p \in C_0 \cup C_j: r \in \sigma_p} w_p \right),$$

while in $(\mathcal{G}, G)$, it holds that

$$\tilde{c}_i(\sigma) = \sum_{r \in \sigma_i} \ell_r(\tilde{n}_{r,i}(\sigma)) = \sum_{r \in \sigma_i} \ell_r \left(\sum_{p \in N: r \in \sigma_p, \{i,p\} \in A} w_p + w_i \right) = \sum_{r \in \sigma_i} \ell_r \left(\sum_{p \in C_0 \cup C_j: r \in \sigma_p} w_p \right),$$

where the last equality follows from the fact that, by construction of G, for any $p \in [n]$ with $p \neq i$, it holds that $\{i, p\} \in A$ if and only if $p \in C_0 \cup C_j$. $\square$

Each game admitting an exact or weighted potential function always admits pure Nash equilibria. Hence, by Proposition 1 and the existence of an exact (resp. weighted) potential function for graphical unweighted (resp. weighted affine) congestion games with undirected social knowledge graphs, we have that d-dimensional unweighted (resp. weighted affine) congestion games always admit pure Nash equilibria.

For weighted affine games, the potential function assume the following expression:

$$\Phi(\sigma) = \sum_{r \in R} \left(\alpha_r \left(\sum_{i \in [n]: r \in \sigma_i} w_i^2 + \sum_{\{i,p\} \in A: r \in \sigma_i \cap \sigma_p} w_i w_p \right) + \beta_r \sum_{i \in [n]: r \in \sigma_i} w_i \right)$$

$$= \frac{1}{2} \sum_{r \in R} \left(\alpha_r \left(\sum_{j=0}^{d} n_{r,j}(\sigma)^2 + \sum_{i \in [n]: r \in \sigma_i} w_i^2 + 2n_{r,0}(\sigma) \sum_{j=1}^{d} n_{r,j}(\sigma) \right) + 2\beta_r \sum_{j=0}^{d} n_{r,j}(\sigma) \right). \quad (1)$$

4. Bounds for the Price of Anarchy

In this section, we provide an upper bound for the price of anarchy of multidimensional weighted affine congestion games as a function of d.

Fix a pure Nash equilibrium σ and a social optimum σ^*, thus fixing the congestions $n_{r,i}(\sigma)$ and $n_{r,i}(\sigma^*)$ for each $i \in [n]$ and $r \in R$. The pure Nash equilibrium condition implies that no player lowers her presumed cost by deviating to the strategy she uses in σ^*. For any player $i \in C_0$, this yields

$$\sum_{r\in\sigma_i} (\alpha_r n_r(\sigma) + \beta_r) - \sum_{r\in\sigma_i^*} (\alpha_r(n_r(\sigma) + w_i) + \beta_r) \le 0, \tag{2}$$

that is a necessary condition satisfied by any pure Nash equilibrium (The equilibrium condition yields the stronger inequality $\sum_{r\in\sigma_i\setminus\sigma_i^*} (\alpha_r n_r(\sigma) + \beta_r) - \sum_{r\in\sigma_i^*\setminus\sigma_i} (\alpha_r(n_r(\sigma) + w_i) + \beta_r) \le 0$, so that inequality $\sum_{r\in\sigma_i} (\alpha_r n_r(\sigma) + \beta_r) - \sum_{r\in\sigma_i^*} (\alpha_r(n_r(\sigma) + w_i) + \beta_r) \le 0$ is a relaxation of the equilibrium condition.). For weighted games, by using $w_i \le n_{r,0}(\sigma^*)$ for any $r \in R$ and $i \in [n]$ such that $r \in \sigma_i$, by multiplying (2) for w_i and summing it for each $i \in C_0$, we get

$$\sum_{r\in R} \left(\alpha_r n_{r,0}(\sigma) \sum_{j=0}^{d} n_{r,j}(\sigma) + \beta_r n_{r,0}(\sigma) - \alpha_r n_{r,0}(\sigma^*) \left(n_{r,0}(\sigma^*) + \sum_{j=0}^{d} n_{r,j}(\sigma) \right) - \beta_r n_{r,0}(\sigma^*) \right) \le 0, \tag{3}$$

that is a further necessary condition satisfied by any pure Nash equilibrium. For unweighted games, we simply fix $w_i = 1$ for any $i \in [n]$ and sum the inequality for each $i \in C_0$, thus getting

$$\sum_{r\in R} \left(\alpha_r n_{r,0}(\sigma) \sum_{j=0}^{d} n_{r,j}(\sigma) + \beta_r n_{r,0}(\sigma) - \alpha_r n_{r,0}(\sigma^*) \left(1 + \sum_{j=0}^{d} n_{r,j}(\sigma) \right) - \beta_r n_{r,0}(\sigma^*) \right) \le 0. \tag{4}$$

For any player $i \in C_j$, with $j \in [d]$, the equilibrium condition yields

$$\sum_{r\in\sigma_i} (\alpha_r (n_{r,j}(\sigma) + n_{r,0}(\sigma)) + \beta_r) - \sum_{r\in\sigma_i^*} (\alpha_r (n_{r,j}(\sigma) + n_{r,0}(\sigma) + w_i) + \beta_r) \le 0.$$

For weighted games, again, by using $w_i \le n_{r,0}(\sigma^*)$ for any $r \in R$ and $i \in [n]$ such that $r \in \sigma_i$, by multiplying this inequality for w_i, and by summing it for each $i \in C_j$, we get

$$\sum_{r\in R} \left(\alpha_r n_{r,j}(\sigma) \left(n_{r,j}(\sigma) + n_{r,0}(\sigma) \right) + \beta_r n_{r,j}(\sigma) - \alpha_r n_{r,j}(\sigma^*) \left(n_{r,j}(\sigma) + n_{r,0}(\sigma) + n_{r,j}(\sigma^*) \right) - \beta_r n_{r,j}(\sigma^*) \right) \le 0.$$

By further summing for each $j \in [d]$, we obtain

$$\sum_{r\in R} \left(\sum_{j=1}^{d} (n_{r,j}(\sigma) (\alpha_r(n_{r,j}(\sigma) + n_{r,0}(\sigma) + \beta_r)) - \sum_{j=1}^{d} (n_{r,j}(\sigma^*) (\alpha_r(n_{r,j}(\sigma) + n_{r,0}(\sigma) + n_{r,j}(\sigma^*)) + \beta_r)) \right) \le 0. \tag{5}$$

For unweighted games, by setting $w_i = 1$, and by summing the equilibrium constraint for any $i \in [n]$ and $j \in [d]$, we analogously get

$$\sum_{r\in R} \left(\sum_{j=1}^{d} (n_{r,j}(\sigma) (\alpha_r(n_{r,j}(\sigma) + n_{r,0}(\sigma) + \beta_r)) - \sum_{j=1}^{d} (n_{r,j}(\sigma^*) (\alpha_r(n_{r,j}(\sigma) + n_{r,0}(\sigma) + 1) + \beta_r)) \right) \le 0. \tag{6}$$

In the sequel, for the sake of conciseness, we adopt $k_{r,j}$ and $l_{r,j}$ as short-hands for $n_{r,j}(\sigma)$ and $n_{r,j}(\sigma^*)$, respectively.

Theorem 1 provides an upper bound for the price of anarchy of multidimensional weighted affine congestion games with respect to social cost function Pres.

Theorem 1. *For each d-dimensional weighted affine congestion game* $(\mathcal{G}, \mathcal{C})$,

$$PoA(\mathcal{G}, \mathcal{C}) \leq \frac{(\sqrt{d+4} + \sqrt{d})(\sqrt{d} \cdot \sqrt{d+4} + d + 4)}{4\sqrt{d+4}} \leq d + 2$$

under the social cost function Pres.

Proof. Let σ and σ^* be a worst-case equilibrium and a social optimum of $(\mathcal{G}, \mathcal{C})$, respectively. Let $k_r = (k_{r,0}, \ldots, k_{r,d})$, $l_r = (l_{r,0}, \ldots, l_{r,d})$, and let $P = (p_{i,j})_{i,j \in [d] \cup \{0\}}$ be the $(d+1) \times (d+1)$ binary matrix such that: (i) $p_{i,j} = 1$ if either $i = j$, or $i = 0$, or $j = 0$; (ii) $p_{i,j} = 0$ otherwise. By summing inequalities (3) and (5), we get the following compact inequality involving the product between vectors, matrices, and scalars:

$$\sum_{r \in R} \left(\alpha_r (k_r \cdot P \cdot k_r^T) + \beta_r \sum_{j=0}^{d} k_{r,j} - \alpha_r (l_r \cdot P \cdot k_r^T + l_r \cdot l_r^T) - \beta_r \sum_{j=0}^{d} l_{r,j} \right) \leq 0 \tag{7}$$

Let $Q = (q_{i,j})_{i,j \in [d] \cup \{0\}}$ be the $(d+1) \times (d+1)$ matrix such that: (i) $q_{i,j} = \sqrt{d}$ if $i = j$; (ii) $q_{i,j} = 1$ if either $i = 0$, or $j = 0$, with $(i,j) \neq (0,0)$; (iii) $q_{i,j} = 0$ otherwise. As $0 \leq p_{i,j} \leq q_{i,j}$ for any i, j we have that

$$l_r \cdot P \cdot k_r^T \leq l_r \cdot Q \cdot k_r^T. \tag{8}$$

We have that matrix Q is a symmetric positive-semidefinite matrix (see Lemma A1 in the Appendix A for the proof of this fact), thus, the following inequality holds for any $u > 0$:

$$0 \leq \left(\sqrt{u} \cdot k_r - \frac{1}{2\sqrt{u}} \cdot l_r \right) \cdot Q \cdot \left(\sqrt{u} \cdot k_r - \frac{1}{2\sqrt{u}} \cdot l_r \right)^T = u \cdot k_r \cdot Q \cdot k_r^T + \frac{1}{4u} \cdot l_r \cdot Q \cdot l_r^T - l_r \cdot Q \cdot k_r^T. \tag{9}$$

Finally, as $0 \leq q_{i,j} \leq \sqrt{d} \cdot p_{i,j}$ for any i, j, we have that

$$x \cdot Q \cdot x^T \leq \sqrt{d} \cdot x \cdot P \cdot x^T \tag{10}$$

for any vector $x = (x_0, \ldots, x_d)$ of non-negative real numbers. By exploiting (7), (9), and (10), for any fixed $u > 0$ we get

$$\begin{aligned}
\text{Pres}(\sigma) &= \sum_{r \in R} \left(\alpha_r (k_r \cdot P \cdot k_r^T) + \beta_r \sum_{j=0}^{d} k_{r,j} \right) \\
&\leq \sum_{r \in R} \left(\alpha_r (l_r \cdot P \cdot k_r^t + l_r \cdot l_r^T) + \beta_r \sum_{j=0}^{d} l_{r,j} \right) \tag{11} \\
&\leq \sum_{r \in R} \left(\alpha_r (l_r \cdot P \cdot k_r^T + l_r \cdot P \cdot l_r^T) + \beta_r \sum_{j=0}^{d} l_{r,j} \right) \\
&\leq \sum_{r \in R} \left(\alpha_r \left(u \cdot k_r \cdot Q \cdot k_r^T + \frac{1}{4u} \cdot l_r \cdot Q \cdot l_r^T + l_r \cdot P \cdot l_r^T \right) + \beta_r \sum_{j=0}^{d} l_{r,j} \right) \tag{12} \\
&\leq \sum_{r \in R} \left(\alpha_r \left(\sqrt{d} \cdot u \cdot k_r \cdot P \cdot k_r^T + \left(\frac{\sqrt{d}}{4u} + 1 \right) \cdot l_r \cdot P \cdot l_r^T \right) + \beta_r \sum_{j=0}^{d} l_{r,j} \right) \tag{13} \\
&\leq \sqrt{d} \cdot u \cdot \sum_{r \in R} \left(\alpha_r (k_r \cdot P \cdot k_r^T) + \beta_r \sum_{j=0}^{d} k_{r,j} \right) + \left(\frac{\sqrt{d}}{4u} + 1 \right) \cdot \sum_{r \in R} \left(\alpha_r (l_r \cdot P \cdot l_r^T) + \beta_r \sum_{j=0}^{d} l_{r,j} \right) \\
&= \sqrt{d} \cdot u \cdot \text{Pres}(\sigma) + \left(\frac{\sqrt{d}}{4u} + 1 \right) \cdot \text{Pres}(\sigma^*), \tag{14}
\end{aligned}$$

where (11), (12), and (13), come from (7), (9), and (10), respectively (Inequalities (11)–(14) can be stated within the smoothness framework of Roughgarden [19], and show that multidimensional weighted affine congestion games are (λ, μ)-smooth with $\lambda = \frac{\sqrt{d}}{4u} + 1$ and $\mu = \sqrt{d} \cdot u$ for any $u > 0$.). Finally, by manipulating (14), we get

$$PoA(\mathcal{G}, \mathcal{C}) = \frac{Pres(\sigma)}{Pres(\sigma^*)} \leq \inf_{u>0} \frac{\frac{\sqrt{d}}{4u} + 1}{1 - \sqrt{d} \cdot u} = \frac{(\sqrt{d+4} + \sqrt{d})(\sqrt{d} \cdot \sqrt{d+4} + d + 4)}{4\sqrt{d+4}}, \tag{15}$$

thus showing the claim. A simpler upper bound of $d + 2$ can be obtained by setting $u = \frac{1}{2\sqrt{d}}$ in (15). □

Relatively to the social cost function Perc, the following upper bound is derived as a corollary of a result in Reference [25].

Corollary 1. *For each d-dimensional affine congestion game $(\mathcal{G}, \mathcal{C})$, $PoA(\mathcal{G}, \mathcal{C}) \leq d(d + 2 + \sqrt{d^2 + 4d})/2$ under the social cost function Perc.*

Proof. Theorem 2 of Reference [25] states that $\delta(G)(\delta(G) + 2 + \sqrt{\delta(G)^2 + 4\delta(G)})/2$ is an upper bound for the price of anarchy of any graphical congestion having independence number $\delta(G)$. As the graphical congestion game equivalent to $(\mathcal{G}, \mathcal{C})$ has independence number equal to d, the claim follows. □

5. Bounds for the Price of Stability

In order to bound the price of stability with respect to the social cost function Pres, we consider a pure Nash equilibrium that minimizes the potential function Φ defined in (1), which leads to the following upper bound.

Theorem 2. *For each d-dimensional weighted affine congestion game $(\mathcal{G}, \mathcal{C})$, $PoS(\mathcal{G}, \mathcal{C}) \leq 2$ under the social cost function Pres.*

Proof. Let σ and σ^* be a pure Nash equilibrium minimizing the potential function Φ defined in (1), and let σ^* be a social optimum. We have that

$$Pres(\sigma) = \sum_{r \in R} \left(\alpha_r \left(\sum_{j=0}^{d} k_{r,j}^2 + 2k_{r,0} \sum_{j=1}^{d} k_{r,j} \right) + \beta_r \sum_{j=0}^{d} k_{r,j} \right)$$

$$\leq \sum_{r \in R} \left(\alpha_r \left(\sum_{j=0}^{d} k_{r,j}^2 + \sum_{i \in [n]: r \in \sigma_i} w_i^2 + 2k_{r,0} \sum_{j=1}^{d} k_{r,j} \right) + 2\beta_r \sum_{j=0}^{d} k_{r,j} \right)$$

$$= 2 \cdot \Phi(\sigma) \tag{16}$$

$$\leq 2 \cdot \Phi(\sigma^*) \tag{17}$$

$$= \sum_{r \in R} \left(\alpha_r \left(\sum_{j=0}^{d} l_{r,j}^2 + \sum_{i \in [n]: r \in \sigma_i^*} w_i^2 + 2l_{r,0} \sum_{j=1}^{d} l_{r,j} \right) + 2\beta_r \sum_{j=0}^{d} l_{r,j} \right) \tag{18}$$

$$\leq \sum_{r \in R} \left(\alpha_r \left(2 \sum_{j=0}^{d} l_{r,j}^2 + 2l_{r,0} \sum_{j=1}^{d} l_{r,j} \right) + 2\beta_r \sum_{j=0}^{d} l_{r,j} \right)$$

$$\leq 2 \sum_{r \in R} \left(\alpha_r \left(\sum_{j=0}^{d} l_{r,j}^2 + 2l_{r,0} \sum_{j=1}^{d} l_{r,j} \right) + \beta_r \sum_{j=0}^{d} l_{r,j} \right)$$

$$= 2 \cdot Pres(\sigma^*), \tag{19}$$

where (17) holds since σ minimizes Φ, and (16) and (18) hold by exploiting (1). By (19), we get $PoS(\mathcal{G},\mathcal{C}) \leq \frac{\text{Pres}(\sigma)}{\text{Pres}(\sigma^*)} \leq 2$, and the claim follows. $\square$

Relatively to the social cost function Perc, the following upper bound is derived as a corollary of a result in Reference [25].

Corollary 2. *For each d-dimensional affine congestion game* $(\mathcal{G},\mathcal{C})$, $PoS(\mathcal{G},\mathcal{C}) \leq 2d$ *under the social cost function* Perc.

Proof. Theorem 6 of Reference [25] states that $2\delta(G)$ is an upper bound for the price of stability of any graphical congestion game having independence number $\delta(G)$. As the graphical congestion game equivalent to $(\mathcal{G},\mathcal{C})$ has independence number equal to d, the claim follows. $\square$

6. Bounds for Bidimensional Unweighted Games

In this section, we investigate in more detail the case of unweighted affine games with $d = 2$, that is, bidimensional affine congestion games, and provide refined bounds for the price of anarchy and the price of stability under both social cost functions. The technique we adopt is the primal-dual framework introduced in Reference [11].

6.1. Price of Anarchy

We first consider the price of anarchy. Let $(\mathcal{G},\mathcal{C})$ be an arbitrary d-dimensional unweighted congestion game, and let σ and σ^* be a worst-case equilibrium and a social optimum of $(\mathcal{G},\mathcal{C})$, respectively. For SF = Pres, we get the following primal linear program $\text{LP}(\text{Pres}, \sigma, \sigma^*)$ in variables $(\alpha_r, \beta_r)_{r \in R}$, whose optimal solution provides an upper bound to $PoA(\mathcal{G},\mathcal{C})$:

$$
\begin{aligned}
max \quad & \sum_{r \in R} \left(\alpha_r \left(\sum_{j=0}^{d} k_{r,j}^2 + 2k_{r,0} \sum_{j=1}^{d} k_{r,j} \right) + \beta_r \sum_{j=0}^{d} k_{r,j} \right) \\
s.t. \quad & \sum_{r \in R} \left(\alpha_r k_{r,0} \sum_{j=0}^{d} k_{r,j} + \beta_r k_{r,0} - \alpha_r l_{r,0} \left(1 + \sum_{j=0}^{d} k_{r,j} \right) - \beta_r l_{r,0} \right) \leq 0 \\
& \sum_{r \in R} \left(\sum_{j=1}^{d} \left(k_{r,j} \left(\alpha_r(k_{r,j} + k_{r,0} + \beta_r) \right) \right) - \sum_{j=1}^{d} \left(l_{r,j} \left(\alpha_r(k_{r,j} + k_{r,0} + 1) + \beta_r \right) \right) \right) \leq 0 \\
& \sum_{r \in R} \left(\alpha_r \left(\sum_{j=0}^{d} l_{r,j}^2 + 2l_{r,0} \sum_{j=1}^{d} l_{r,j} \right) + \beta_r \sum_{j=0}^{d} l_{r,j} \right) = 1 \\
& \alpha_r, \beta_r \geq 0 \hspace{6cm} \forall r \in R.
\end{aligned}
$$

The optimal solution of the above linear program is an upper bound to the price of anarchy as the objective function is equal to $\text{Pres}(\sigma)$, the first two constraints are the pure Nash equilibrium conditions derived in (4) and (6), respectively (which are necessary conditions satisfied by any equilibrium), and the last normalization constraint imposes without loss of generality that $\text{Pres}(\sigma^*) = 1$ (When applying the primal dual method, we observe that, once σ and σ^* are fixed, the coefficients $(\alpha_r)_{r \in R}$ and $(\beta_r)_{r \in R}$ are chosen in such a way that the value $\text{Pres}(\sigma) = \text{Pres}(\sigma)/\text{Pres}(\sigma^*)$ is maximized, thus getting an upper bound on the price of anarchy. We also observe that $(\alpha_r)_{r \in R}$ and $(\beta_r)_{r \in R}$ are the unique variables in the considered LP formulation, and the other quantities (e.g., the congestions) are considered as fixed parameters (w.r.t. the LP formulation). See Reference [11] for further details on the primal-dual method and how to apply it to measure the performance of congestion games under different quality metrics.).

By associating the three dual variables x, y and γ, with the three constraints of $LP(\text{Pres}, \sigma, \sigma^*)$, the dual formulation $DLP(\text{Pres}, \sigma, \sigma^*)$ becomes

$$\min \quad \gamma$$

s.t.

$$x \left(k_{r,0} \sum_{j=0}^{d} k_{r,j} - l_{r,0} - l_{r,0} \sum_{j=0}^{d} k_{r,j} \right) + y \sum_{j=1}^{d} (k_{r,j}(k_{r,j} + k_{r,0}) - l_{r,j}(k_{r,j} + k_{r,0} + 1))$$

$$+ \gamma \left(\sum_{j=0}^{d} l_{r,j}^2 + 2l_{r,0} \sum_{j=1}^{d} l_{r,j} \right) \geq \sum_{j=0}^{d} k_{r,j}^2 + 2k_{r,0} \sum_{j=1}^{d} k_{r,j} \qquad \forall r \in R$$

$$x(k_{r,0} - l_{r,0}) + y \sum_{j=1}^{d} (k_{r,j} - l_{r,j}) + \gamma \sum_{j=0}^{d} l_{r,j} \geq \sum_{j=0}^{d} k_{r,j} \qquad \forall r \in R$$

$$x, y \geq 0.$$

By the Weak Duality Theorem, each feasible solution for $DLP(\text{Pres}, \sigma, \sigma^*)$ provides an upper bound to the optimal solution of $LP(\text{Pres}, \sigma, \sigma^*)$, that is on the price of anarchy achievable by the particular choice of σ and σ^*. Anyway, if the provided dual solution is independent of this choice, we obtain an upper bound on the price of anarchy for any possible game.

For the case of the social cost function Perc, we only need to replace the objective function and the third constraint in $LP(\text{Pres}, \sigma, \sigma^*)$, respectively, with

$$\sum_{r \in R} \left(\alpha_r \left(\sum_{j=0}^{d} k_{r,j} \right)^2 + \beta_r \sum_{j=0}^{d} k_{r,j} \right) \quad \text{and} \quad \sum_{r \in R} \left(\alpha_r \left(\sum_{j=0}^{d} l_{r,j} \right)^2 + \beta_r \sum_{j=0}^{d} l_{r,j} \right) = 1.$$

This results in the following dual program $DLP(\text{Perc}, \sigma, \sigma^*)$:

$$\min \quad \gamma$$

s.t.

$$x \left(k_{r,0} \sum_{j=0}^{d} k_{r,j} - l_{r,0} - l_{r,0} \sum_{j=0}^{d} k_{r,j} \right) + y \sum_{j=1}^{d} (k_{r,j}(k_{r,j} + k_{r,0}) - l_{r,j}(k_{r,j} + k_{r,0} + 1))$$

$$+ \gamma \left(\sum_{j=0}^{d} l_{r,j} \right)^2 \geq \left(\sum_{j=0}^{d} k_{r,j} \right)^2 \qquad \forall r \in R$$

$$x(k_{r,0} - l_{r,0}) + y \sum_{j=1}^{d} (k_{r,j} - l_{r,j}) + \gamma \sum_{j=0}^{d} l_{r,j} \geq \sum_{j=0}^{d} k_{r,j} \qquad \forall r \in R$$

$$x, y \geq 0.$$

Note that the second constraint is the same in both $DLP(\text{Pres}, \sigma, \sigma^*)$ and $DLP(\text{Perc}, \sigma, \sigma^*)$. For the sake of conciseness, in the sequel, we shall drop the subscript r from the notation; moreover, when fixed a dual solution, we shall denote the first and second constraint of a given dual program as $g_1(k, l) \geq 0$ and $g_2(k, l) \geq 0$, respectively, where we set $k = (k_0, \dots, k_d)$ and $l = (l_0, \dots, l_d)$.

When $d = 2$, we exploit an equivalent, but nicer, representation of the dual inequalities. With this aim, we set $k_r := n_r(\sigma)$ and $l_r := n_r(\sigma^*)$ and replace $k_{r,0}$ and $l_{r,0}$ with $k_r - k_{r,1} - k_{r,2}$ and $l_r - l_{r,1} - l_{r,2}$, respectively. By substituting and rearranging, $DLP(\text{Pres}, \sigma, \sigma^*)$ becomes

$$
\begin{aligned}
min \quad & \gamma \\
s.t. \quad &
\end{aligned}
$$

$$
x\left((k_r - k_{r,1} - k_{r,2})k_r - (l_r - l_{r,1} - l_{r,2})(k_r + 1)\right)
$$
$$
+ y\left(k_{r,1}(k_r - k_{r,2}) - l_{r,1}(k_r - k_{r,2} + 1) + k_{r,2}(k_r - k_{r,1}) - l_{r,2}(k_r - k_{r,1} + 1)\right)
$$
$$
+ \gamma\left(l_r^2 - 2l_{r,1}l_{r,2}\right) \geq k_r^2 - 2k_{r,1}k_{r,2} \qquad \forall r \in R
$$
$$
x(k_r - k_{r,1} - k_{r,2} - l_r + l_{r,1} + l_{r,2}) + y(k_{r,1} + k_{r,2} - l_{r,1} - l_{r,2}) + \gamma l_r \geq k_r \qquad \forall r \in R
$$
$$
x, y \geq 0.
$$

Similarly, the dual program DLP(Perc, σ, σ^*) can be rewritten as:

$$
\begin{aligned}
min \quad & \gamma \\
s.t. \quad &
\end{aligned}
$$

$$
x\left((k_r - k_{r,1} - k_{r,2})k_r - (l_r - l_{r,1} - l_{r,2})(k_r + 1)\right)
$$
$$
+ y\left(k_{r,1}(k_r - k_{r,2}) - l_{r,1}(k_r - k_{r,2} + 1) + k_{r,2}(k_r - k_{r,1}) - l_{r,2}(k_r - k_{r,1} + 1)\right) + \gamma l_r^2 \geq k_r^2 \qquad \forall r \in R
$$
$$
x(k_r - k_{r,1} - k_{r,2} - l_r + l_{r,1} + l_{r,2}) + y(k_{r,1} + k_{r,2} - l_{r,1} - l_{r,2}) + \gamma l_r \geq k_r \qquad \forall r \in R
$$
$$
x, y \geq 0.
$$

In the following theorem we provide upper bounds for the price of anarchy of bidimensional affine congestion games under social cost functions Pres and Perc.

Theorem 3. *For each bidimensional affine congestion game* $(\mathcal{G}, \mathcal{C})$*,* $PoA(\mathcal{G}, \mathcal{C}) \leq \frac{119}{33}$ *under the social cost function* Pres *and* $PoA(\mathcal{G}, \mathcal{C}) \leq \frac{35}{8}$ *under the social cost function* Perc*.*

We now show the existence of two matching lower bounding instances (the proof is deferred to the Appendix B).

Theorem 4. *There exist two bidimensional linear congestion games* $(\mathcal{G}, \mathcal{C})$ *and* $(\mathcal{G}', \mathcal{C}')$ *such that* $PoA(\mathcal{G}, \mathcal{C}) \geq \frac{119}{33}$ *under the social cost function* Pres *and* $PoA(\mathcal{G}', \mathcal{C}') \geq \frac{35}{8}$ *under the social cost function* Perc *Appendix B.*

6.2. Price of Stability

In order to bound the price of stability, we can use the same primal formulations exploited for the determination of the price of anarchy with the additional constraint $\Phi(\sigma) \leq \Phi(\sigma^*)$, which, by Equation (1), becomes

$$
\sum_{r \in R} \left(\alpha_r \left(\sum_{j=0}^{d} \left(k_{r,j}^2 + k_{r,j} - l_{r,j}^2 - l_{r,j} \right) + 2k_{r,0} \sum_{j=1}^{d} k_{r,j} - 2l_{r,0} \sum_{j=1}^{d} l_{r,j} \right) + 2\beta_r \sum_{j=0}^{d} (k_{r,j} - l_{r,j}) \right) \leq 0.
$$

Hence, the dual program for the social cost function Pres becomes the following one.

$$min \quad \gamma$$
s.t.

$$x\left(k_{r,0}\sum_{j=0}^{d}k_{r,j} - l_{r,0} - l_{r,0}\sum_{j=0}^{d}k_{r,j}\right) + y\sum_{j=1}^{d}\left(k_{r,j}(k_{r,j}+k_{r,0}) - l_{r,j}(k_{r,j}+k_{r,0}+1)\right)$$

$$+z\left(\sum_{j=0}^{d}\left(k_{r,j}^2 + k_{r,j} - l_{r,j}^2 - l_{r,j}\right) + 2k_{r,0}\sum_{j=1}^{d}k_{r,j} - 2l_{r,0}\sum_{j=1}^{d}l_{r,j}\right)$$

$$+\gamma\left(\sum_{j=0}^{d}l_{r,j}^2 + 2l_{r,0}\sum_{j=1}^{d}l_{r,j}\right) \geq \sum_{j=0}^{d}k_{r,j}^2 + 2k_{r,0}\sum_{j=1}^{d}k_{r,j} \qquad \forall r \in R$$

$$x(k_{r,0}-l_{r,0}) + y\sum_{j=1}^{d}(k_{r,j}-l_{r,j}) + 2z\sum_{j=0}^{d}(k_{r,j}-l_{r,j}) + \gamma\sum_{j=0}^{d}l_{r,j} \geq \sum_{j=0}^{d}k_{r,j} \qquad \forall r \in R$$

$$x, y, z \geq 0.$$

Again, for the social cost function Perc, we obtain mutatis mutandis the following dual program.

$$min \quad \gamma$$
s.t.

$$x\left(k_{r,0}\sum_{j=0}^{d}k_{r,j} - l_{r,0} - l_{r,0}\sum_{j=0}^{d}k_{r,j}\right) + y\sum_{j=1}^{d}\left(k_{r,j}(k_{r,j}+k_{r,0}) - l_{r,j}(k_{r,j}+k_{r,0}+1)\right)$$

$$+z\left(\sum_{j=0}^{d}\left(k_{r,j}^2 + k_{r,j} - l_{r,j}^2 - l_{r,j}\right) + 2k_{r,0}\sum_{j=1}^{d}k_{r,j} - 2l_{r,0}\sum_{j=1}^{d}l_{r,j}\right)$$

$$+\gamma\left(\sum_{j=0}^{d}l_{r,j}\right)^2 \geq \left(\sum_{j=0}^{d}k_{r,j}\right)^2 \qquad \forall r \in R$$

$$x(k_{r,0}-l_{r,0}) + y\sum_{j=1}^{d}(k_{r,j}-l_{r,j}) + 2z\sum_{j=0}^{d}(k_{r,j}-l_{r,j}) + \gamma\sum_{j=0}^{d}l_{r,j} \geq \sum_{j=0}^{d}k_{r,j} \qquad \forall r \in R$$

$$x, y, z \geq 0.$$

Again, by setting $k_r := n_r(\sigma)$ and $l_r := n_r(\sigma^*)$ and replacing $k_{r,0}$ and $l_{r,0}$ with $k_r - k_{r,1} - k_{r,2}$ and $l_r - l_{r,1} - l_{r,2}$, respectively, DLP(Pres, σ, σ^*) becomes

$$min \quad \gamma$$
s.t.

$$x\left((k_r - k_{r,1} - k_{r,2})k_r - (l_r - l_{r,1} - l_{r,2})(k_r + 1)\right)$$
$$+y\left(k_{r,1}(k_r - k_{r,2}) - l_{r,1}(k_r - k_{r,2}+1) + k_{r,2}(k_r - k_{r,1}) - l_{r,2}(k_r - k_{r,1}+1)\right)$$
$$+z\left(k_r^2 - 2k_{r,1}k_{r,2} - l_r^2 + 2l_{r,1}l_{r,2} + k_r - l_r\right) + \gamma\left(l_r^2 - 2l_{r,1}l_{r,2}\right) \geq k_r^2 - 2k_{r,1}k_{r,2} \qquad \forall r \in R$$
$$x(k_r - k_{r,1} - k_{r,2} - l_r + l_{r,1} + l_{r,2}) + y(k_{r,1} + k_{r,2} - l_{r,1} - l_{r,2}) + 2z(k_r - l_r) + \gamma l_r \geq k_r \qquad \forall r \in R$$
$$x, y, z \geq 0.$$

Similarly, the dual program DLP(Perc, σ, σ^*) can be rewritten as:

$$min \quad \gamma$$
s.t.

$$x\left((k_r - k_{r,1} - k_{r,2})k_r - (l_r - l_{r,1} - l_{r,2})(k_r + 1)\right)$$
$$+y\left(k_{r,1}(k_r - k_{r,2}) - l_{r,1}(k_r - k_{r,2}+1) + k_{r,2}(k_r - k_{r,1}) - l_{r,2}(k_r - k_{r,1}+1)\right)$$
$$+z\left(k_r^2 - 2k_{r,1}k_{r,2} - l_r^2 + 2l_{r,1}l_{r,2} + k_r - l_r\right) + \gamma l_r^2 \geq k_r^2 \qquad \forall r \in R$$
$$x(k_r - k_{r,1} - k_{r,2} - l_r + l_{r,1} + l_{r,2}) + y(k_{r,1} + k_{r,2} - l_{r,1} - l_{r,2}) + 2z(k_r - l_r) + \gamma l_r \geq k_r \qquad \forall r \in R$$
$$x, y, z \geq 0.$$

Theorem 5. *For each bidimensional affine congestion game* $(\mathcal{G},\mathcal{C})$, $PoS(\mathcal{G},\mathcal{C}) \leq 1 + \frac{2}{\sqrt{7}}$ *under the social cost function* Pres *and* $PoS(\mathcal{G},\mathcal{C}) \leq 2.92$ *under the social cost function* Perc.

Proof. For the social cost function Pres, set $\gamma = 1 + \frac{2}{\sqrt{7}}$, $x = y = \frac{1}{\sqrt{7}}$ and $z = \frac{1}{2} + \frac{1}{2\sqrt{7}}$. The second dual constraint is always satisfied, as $\min\{x,y\} \geq 1$ and $\max\{x,y\} + 2z \leq \gamma$. Thus, we shall focus again on the first constraint $g_1(k,l) \geq 0$. For any $r \in R$, $g_1(k,l)$ becomes

$$k^2(3-\sqrt{7}) - k(2l-1-\sqrt{7}) + 2k_1k_2(\sqrt{7}-3) + 2(k_1l_2+k_2l_1) + (l^2-l)(3+\sqrt{7}) - 2l_1l_2(3+\sqrt{7}) \geq 0.$$

The claim follows by applying Lemma A9 reported in the Appendix A.

For the social cost function Perc, set $\gamma = 2.92$, $x = 0.68$, $y = 1.3$ and $z = 0.81$. Again, the second dual constraint is always satisfied, as $\min\{x,y\} \geq 1$ and $\max\{x,y\} + 2z \leq \gamma$. Thus, we shall focus again on the first constraint $g_1(k,l) \geq 0$. For any $r \in R$, $g_1(k,l)$ become $49k^2 + k(62k_1 + 62k_2 - 68l - 62l_1 - 62l_2 + 81) + 130k_1l_2 + 130k_2l_1 - 422k_1k_2 + 211l^2 - 149l + 2(81l_1l_2 - 31l_1 - 31l_2) \geq 0$. The claim follows by applying Lemma A13 reported in the Appendix A. □

For these cases, unfortunately, we are not able to devise matching lower bounds. The following result is obtained by suitably extending the lower bounding instance given in Reference [17] for the price of stability of congestion games (the proof is deferred to the Appendix).

Theorem 6. *For any* $\epsilon > 0$, *there exist two bidimensional linear congestion games* $(\mathcal{G},\mathcal{C})$ *and* $(\mathcal{G}',\mathcal{C}')$ *such that* $PoS(\mathcal{G},\mathcal{C}) \geq \frac{1+\sqrt{5}}{2} - \epsilon$ *under the social cost function* Pres *and* $PoS(\mathcal{G}',\mathcal{C}') \geq \frac{5+\sqrt{17}}{4} - \epsilon$ *under the social cost function* Perc.

7. Conclusions and Open Problems

We have introduced d-dimensional (weighted) congestion games: a generalization of (weighted) congestion games able to model various interesting scenarios of applications. They can also be reinterpreted as a particular subclass of that of graphical (weighted) congestion games defined by an undirected social knowledge graph whose independence number is equal to d. We have provided bounds for the price of anarchy and the price of stability of these games as a function of d under the two fundamental social cost functions sum of the players' perceived costs and sum of the players' presumed costs. We have also considered in deeper detail the case of $d = 2$ in presence of unweighted players only.

Closing the gap between upper and lower bounds is an intriguing and challenging open problem. In particular, we conjecture that the upper bound of $O(d)$ for the price of anarchy of d-dimensional weighted congestion games is asymptotically tight (with respect to d), even for unweighted games.

Along the line of research of improving the performance of congestion games via some feasible strategies or coordination (e.g., taxes [27,28] or Stackelberg strategies [29,30]), another interesting research direction is partitioning the players into $d + 1$ clusters similarly as in d-dimensional games, to improve as much as possible the price of anarchy or the price of stability.

A further research direction is that of combining the model of multidimensional congestion games with other variants of congestion games (e.g., risk-averse congestion games [31–34] and congestion games with link failures [35–37]).

Author Contributions: Conceptualization, V.B., M.F., V.G., and C.V.; Methodology, V.B., M.F., V.G., and C.V.; Validation, V.B., M.F., and C.V.; Formal Analysis, V.B., M.F., and C.V.; Investigation, V.B., M.F., V.G., and C.V.; Writing Original Draft Preparation, V.B., M.F., and C.V.; Writing Review & Editing, V.B. and C.V.; Visualization, V.B., M.F., V.G., and C.V.; Supervision, V.B., M.F., and C.V.; Project Administration, V.B. and M.F.; Funding Acquisition, M.F. All authors have read and agreed to the published version of the manuscript.

Funding: This work was partially supported by the Italian MIUR PRIN 2017 Project ALGADIMAR "Algorithms, Games, and Digital Markets".

Conflicts of Interest: The authors declare no conflict of interest.

Appendix A. Technical Lemmas

In this section we gather all technical lemmas needed to prove our main theorems.

Lemma A1. *For any $d \geq 0$, let $Q = (q_{i,j})_{i,j \in [d] \cup \{0\}}$ be the $(d+1) \times (d+1)$ matrix such that: (i) $q_{i,j} = \sqrt{d}$ if $i = j$; (ii) $q_{i,j} = 1$ if either $i = 0$, or $j = 0$, with $(i,j) \neq (0,0)$; (iii) $q_{i,j} = 0$ otherwise. We have that Q is a positive-semidefinite matrix.*

Proof. To show the claim, we resort to the Sylvester's criterion, stating that a symmetric matrix M is positive-semidefinite if and only if the determinant of each principal minor of M (i.e., each upper upper left h-by-h corner of M) is non-negative. Let $A_{h,x} = (a_{h,x,i,j})_{i,j \in [h]}$ be a $h \times h$ matrix such that: (i) $a_{h,x,i,j} = x$ if $i = j$; (ii) $a_{h,x,i,j} = 1$ if $(i,j) \neq (1,1)$, and, either $i = 1$, or $j = 1$; (iii) $a_{h,x,i,j} = 0$ otherwise. We have that each principal minor of matrix Q is of type $A_{h,\sqrt{d}}$ for some $h \in [d+1]$. Thus, it is sufficient showing that the determinant of matrix $A_{h,\sqrt{d}}$, denoted as $Det(A_{h,\sqrt{d}})$, is non-negative for any $h \in [d+1]$.

We first show by induction on integers $h \geq 1$ that $Det(A_{h,x}) = x^h - (h-1) \cdot x^{h-2}$ for any fixed $x \in \mathbb{R}$. If $h = 0$ we trivially get $Det(A_{h,x}) = x = x^h - (h-1) \cdot x^{h-2}$. Now, we assume that $Det(A_{h,x}) = x^h - (h-1)x^{h-2}$ holds for some $h \geq 1$, and we show that $Det(A_{h+1,x}) = x^{h+1} - h \cdot x^{h-1}$. We get $Det(A_{h+1,x}) = x \cdot Det(A_{h,x}) - x^{h-1} = x(x^h - (h-1)x^{h-2}) - x^{h-1} = x^{h+1} - h \cdot x^{h-1}$, where the first equality comes from the Laplace expansion for computing the determinant, and the second equality comes from the inductive hypothesis.

By using the fact that $Det(A_{h,x}) = x^h - (h-1) \cdot x^{h-2}$ holds for any $x \in \mathbb{R}$ and any integer $h \geq 1$, we have that $Det(A_{h,\sqrt{d}}) = (\sqrt{d})^h - (h-1)(\sqrt{d})^{h-2} \geq 0$ for any $h \in [d+1]$, where the last inequality holds since quantity $x^h - (h-1)x^{h-1}$ is always non-negative for any $x \geq \sqrt{h-1}$ if $h \leq d+1$. Thus each principal minor of Q has a non-negative determinant, and the claim follows. $\square$

Lemma A2. *Let $\theta : \mathbb{Z}_{\geq 0}^6 \to \mathbb{Q}$ be the function such that $\theta(a,b,c,d,e,f) = 18a^2 - a(b+c+51d - e - f) + 50bf + 50ce - 34bc + 119d^2 - 51d + e + f - 238ef$. For any $(a,b,c,d,e,f) \in \mathbb{Z}_{\geq 0}^6$ such that $a \geq b + c$ and $d \geq e + f$, it holds that $\theta(a,b,c,d,e,f) \geq 0$.*

Proof. At a first glance, in order to use standard arguments from calculus, we allow the 6-tuples (a,b,c,d,e,f) to take values in the set of non-negative real numbers.

We first show that, in such an extended scenario, θ attains its minimum for 6-tuples (a,b,c,d,e,f) such that $b = c$ and $e = f$. Consider to this aim the 6-tuple $(a,b,b+h,d,e,e+k)$, where $h,k \in \mathbb{R}$. By definition of θ, we get $\theta(a, b + \frac{h}{2}, b + \frac{h}{2}, d, e + \frac{k}{2}, e + \frac{k}{2}) = \theta(a,b,b+h,d,e,e+k) - \frac{17h^2 - 50hk + 119k^2}{2} \leq \theta(a,b,b+h,d,e,e+k) - \frac{(4h-10k)^2}{2} \leq \theta(a,b,b+h,d,e,e+k)$.

Hence, we do not lose in generality by restricting to 6-tuples of non-negative real values (a,b,b,d,e,e) such that $a \geq 2b$ and $d \geq 2e$. In this case θ becomes $18a^2 - a(2b + 51d - 2e) - 34b^2 + 100be + 119d^2 - 51d - 238e^2 + 2e$. Consider the two partial derivatives $\frac{\delta\theta}{\delta b} = 100e - 2a - 68b$ and $\frac{\delta\theta}{\delta e} = 2(a + 50b + 1 - 238e)$. Since they are linear and decreasing in b and e, respectively, it follows that θ is minimized at one of the following four cases: $b = 0 \wedge e = 0$, $b = 0 \wedge e = \frac{d}{2}$, $b = \frac{a}{2} \wedge e = 0$ and $b = \frac{a}{2} \wedge e = \frac{d}{2}$.

In the first case, θ becomes $18a^2 - 51ad + 119d^2 - 51d$. Since $\frac{\delta\theta}{\delta a} = 36a - 51d$, θ is minimized at $a = \frac{17d}{12}$. By substituting, θ becomes $\frac{1}{8}(663d^2 - 408d)$ which is always non-negative for any $d \in \mathbb{Z}$.

In the second case, θ becomes $36a^2 - 100ad + 119d^2 - 100d$. Since $\frac{\delta\theta}{\delta a} = 36a - 50d$, θ is minimized at $a = \frac{25d}{18}$. By substituting, θ becomes $\frac{1}{9}(223d^2 - 450d)$ which is always non-negative for any $d \in \mathbb{Z} \setminus \{1,2\}$.

In the third case, θ becomes $\frac{17}{2}(a^2 - 6ad + 14d^2 - 6d)$. Since $\frac{\delta\theta}{\delta a} = 17(a - 3d)$, θ is minimized at $a = 3d$. By substituting, θ becomes $\frac{17}{2}(5d^2 - 6d)$ which is always non-negative for any $d \in \mathbb{Z} \setminus \{1\}$.

In the fourth case, θ becomes $\frac{1}{2}(17a^2 - 50ad + 119d^2 - 100d)$. Since $\frac{\delta\theta}{\delta a} = 17a - 25d$, θ is minimized at $a = \frac{25d}{17}$. By substituting, θ becomes $\frac{1}{17}(699d^2 - 850d)$ which is always non-negative for any $d \in \mathbb{Z} \setminus \{1\}$.

Hence, in order to complete the proof, we are left to settle the following cases: $(a, 0, 0, 1, 0, 0)$, $(a, 0, 0, 2, 1, 1)$, $(a, 0, 0, 1, 1, 0)$, $(a, 0, 0, 1, 0, 1)$, $(a, \frac{a}{2}, \frac{a}{2}, 1, 1, 0)$, $(a, \frac{a}{2}, \frac{a}{2}, 1, 0, 1)$ and $(a, \frac{a}{2}, \frac{a}{2}, 1, 0, 0)$.

In the case $(a, 0, 0, 1, 0, 0)$, θ becomes $18a^2 - 51a + 68$ which is always non-negative for any $a \in \mathbb{R}$. In the case $(a, 0, 0, 2, 1, 1)$, θ becomes $18a^2 - 100a + 138$ which is always non-negative for any $a \in \mathbb{Z}$. In the cases $(a, 0, 0, 1, 1, 0)$ and $(a, 0, 0, 1, 0, 1)$, θ becomes $18a^2 - 50a + 69$ which is always non-negative for any $a \in \mathbb{R}$. In the cases $(a, \frac{a}{2}, \frac{a}{2}, 1, 1, 0)$ and $(a, \frac{a}{2}, \frac{a}{2}, 1, 0, 1)$, θ becomes $\frac{17a^2 - 50a + 138}{2}$ which is always non-negative for any $a \in \mathbb{R}$. Finally, in the case $(a, \frac{a}{2}, \frac{a}{2}, 1, 0, 0)$, θ becomes $\frac{17}{2}(a^2 - 6a + 8)$ which is always non-negative for any $a \in \mathbb{Z} \setminus \{3\}$. Hence, we are only left to consider the case $(3, b, c, 1, 0, 0)$ for which θ becomes $77 - 34bc - 3(b + c)$. Since $b + c \leq 3$, it holds that $77 - 34bc - 3(b + c) \geq 68 - 34bc$ which is always non-negative since $bc \leq 2$ for any $b, c \in \mathbb{Z}_{\geq 0}$ such that $b + c \leq 3$. $\square$

Lemma A3. *Let* $\lambda : \mathbb{Z}_{\geq 0}^2 \to \mathbb{Q}$ *be the function such that* $\lambda(a, d) = a^2 - 3ad + 5d^2 - 3d$. *For any* $(a, d) \in \mathbb{Z}_{\geq 0}^2$ *such that* $d \neq 1$, *it holds that* $\lambda(a, d) \geq 0$.

Proof. Since $\frac{\delta\lambda}{\delta a} = 2a - 3d$, λ is minimized at $a = \frac{3}{2}d$. By substituting, we get $11d^2 - 12d$ which is non-negative for any $d \in \mathbb{Z} \setminus \{1\}$. $\square$

Lemma A4. *Let* $\lambda : \mathbb{Z}_{\geq 0}^2 \to \mathbb{Q}$ *be the function such that* $\lambda(a, d) = a^2 - 6ad + 14d^2 - 6d$. *For any* $(a, d) \in \mathbb{Z}_{\geq 0}^2$ *such that* $d \neq 1$, *it holds that* $\lambda(a, d) \geq 0$.

Proof. Since $\frac{\delta\lambda}{\delta a} = 2a - 6d$, λ is minimized at $a = 3d$. By substituting, we get $5d^2 - 6d$ which is non-negative for any $d \in \mathbb{Z} \setminus \{1\}$. $\square$

Lemma A5. *Let* $\lambda : \mathbb{Z}_{\geq 0}^2 \to \mathbb{Q}$ *be the function such that* $\lambda(a, d) = 20a^2 - 84ad + 259d^2 - 168d$. *For any* $(a, d) \in \mathbb{Z}_{\geq 0}^2$, *it holds that* $\lambda(a, d) \geq 0$.

Proof. Since $\frac{\delta\lambda}{\delta a} = 40a - 84d$, λ is minimized at $a = \frac{21}{10}d$. By substituting, we get $61d^2 - 60d$ which is non-negative for any $d \in \mathbb{Z}$. $\square$

Lemma A6. *Let* $\lambda : \mathbb{Z}_{\geq 0}^2 \to \mathbb{Q}$ *be the function such that* $\lambda(a, d) = 13a^2 - 21ad + 35d^2 - 21d$. *For any* $(a, d) \in \mathbb{Z}_{\geq 0}^2$, *it holds that* $\lambda(a, d) \geq 0$.

Proof. Since $\frac{\delta\lambda}{\delta a} = 26a - 21d$, λ is minimized at $a = \frac{21}{26}d$. By substituting, we get $197d^2 - 156d$ which is non-negative for any $d \in \mathbb{Z}$. $\square$

Lemma A7. *Let* $\theta : \mathbb{Z}_{\geq 0}^6 \to \mathbb{Q}$ *be the function such that* $\theta(a, b, c, d, e, f) = 7a^2 + 3a(2b + 2c - 5d - 2e - 2f) + 21bf + 21ce - 42bc + 35d^2 - 15d - 6e - 6f$. *For any* $(a, b, c, d, e, f) \in \mathbb{Z}_{\geq 0}^6$ *such that* $a \geq b + c$ *and* $d \geq e + f$, *it holds that* $\theta(a, b, c, d, e, f) \geq 0$.

Proof. At a first glance, in order to use standard arguments from calculus, we allow the 6-tuples (a, b, c, d, e, f) to take values in the set of non-negative real numbers. Since $\frac{\delta\theta}{\delta c} = 3(2a - 14b + 7e)$ and $\frac{\delta\theta}{\delta f} = 3(7b - 2a - 2)$, θ is minimized at one of the following four cases: $c = 0 \wedge f = 0$, $c = 0 \wedge f = d - e$, $c = a - b \wedge f = 0$ and $c = a - b \wedge f = d - e$.

In the first case, we get $\theta = 7a^2 + 3a(2b - 5d - 2e) + 35d^2 - 15d - 6e$. Since $\frac{\delta\theta}{\delta b} = 6a$, θ is minimized at $b = 0$ which yields $\theta = 7a^2 - 3a(5d + 2e) + 35d^2 - 15d - 6e$. Since $\frac{\delta\theta}{\delta e} = -6(a + 1)$, θ is minimized at $e = d$ which yields $\theta = 7(a^2 - 3ad + 5d^2 - 3d)$. The claim then follows for any $d \neq 1$ by applying Lemma A3. For the leftover tuples of the form $(a, 0, 0, 1, 1, 0)$, we get $\theta = 7(a^2 - 3a + 2)$ which is always non-negative for any $a \in \mathbb{Z}$.

In the second case, we get $\theta = 7a^2 + 3a(2b - 7d) + 7(3b(d - e) + 5d^2 - 3d)$. Since $\frac{\delta\theta}{\delta b} = 3(2a + 7(d - e))$, θ is minimized at $b = 0$, which yields $\theta = 7(a^2 - 3ad + 5d^2 - 3d)$. The claim then follows for any $d \neq 1$ by applying Lemma A3. For the leftover tuples of the form $(a, 0, 0, 1, e, 1 - e)$, we get $\theta = 7(a^2 - 3a + 2)$ which is always non-negative for any $a \in \mathbb{Z}$.

In the third case, we get $\theta = 13a^2 - 3a(14b + 5(d - e)) + 42b^2 - 21be + 35d^2 - 15d - 6e$. Since $\frac{\delta\theta}{\delta e} = 3(5a - 7b - 2)$, θ is minimized at either $e = 0$ or $e = d$. For $e = d$, we get $\theta = 13a^2 - 42ab + 42b^2 - 21bd + 35d^2 - 21d$. Since $\frac{\delta\theta}{\delta b} = -21(2a - 4b + d)$, θ is minimized at $b = \frac{2a+d}{4}$. This yields $\theta = 20a^2 - 84ad + 259d^2 - 168d$ and the claim then follows by applying Lemma A5. For $e = 0$, we get $\theta = 13a^2 - 3a(14b + 5d) + 42b^2 + 35d^2 - 15d$. Since $\frac{\delta\theta}{\delta b} = 42(2b - a)$, θ is minimized at $b = \frac{a}{2}$ which yields $\theta = 5(a^2 - 6ad + 14d^2 - 6d)$ and the claim then follows for any $d \neq 1$ by applying Lemma A4. For the leftover tuples of the form $(a, \frac{a}{2}, \frac{a}{2}, 1, 0, 0)$, we get $\theta = \frac{5}{2}(a^2 - 6a + 8)$ which is always non-negative for any $a \in \mathbb{Z} \setminus \{3\}$. Hence, we are still left to prove what happens for the tuples of the form $(3, b, 3 - b, 1, 0, 0)$. In this case, we get $\theta = 42b^2 - 126b + 92$ which is always non-negative for any $b \in \mathbb{Z}$.

In the fourth case, we get $\theta = 13a^2 - 21a(2b + d - e) + 7(6b^2 + 3b(d - 2e) + 5d^2 - 3d)$. Since $\frac{\delta\theta}{\delta e} = 21(a - 2b)$, θ is minimized at either $e = 0$ or $e = d$. For $e = 0$, we get $\theta = 13a^2 - 21a(2b + d) + 7(6b^2 + 3bd + 5d^2 - 3d)$. Since $\frac{\delta\theta}{\delta b} = -21(2a - 4b - d)$, θ is minimized at either $b = 0$ or $b = \frac{2a-d}{4}$. The first case yields $\theta = 13a^2 - 21ad + 35d^2 - 21d$ and the claim then follows by applying Lemma A6, while the second one yields $\theta = \frac{20a^2 - 84ad + 259d^2 - 168d}{8}$ and the claim then follows by applying Lemma A5. For $e = d$, we get $\theta = 13a^2 - 41ab + 7(6b^2 - 3bd + 5d^2 - 3d)$. Since $\frac{\delta\theta}{\delta b} = -21(2a - 4b + d)$, θ is minimized at $b = \frac{2a+d}{4}$ which yields $\theta = \frac{20a^2 - 84ad + 259d^2 - 168d}{8}$ and the claim then follows by applying Lemma A5. □

Lemma A8. *Let* $\lambda : \mathbb{Z}^2_{\geq 0} \to \mathbb{Q}$ *be the function such that* $\lambda(a, d) = \frac{3-\sqrt{7}}{2}a^2 + (1 + \sqrt{7} - 2d)a + (3 + \sqrt{7})(\frac{d^2}{2} - d)$. *For any* $(a, d) \in \mathbb{Z}^2_{\geq 0} \setminus \{(0, 1), (1, 1), (1, 2)\}$, *it holds that* $\lambda(a, d) \geq 0$.

Proof. Since $\frac{\delta\lambda}{\delta a} = (3 + \sqrt{7})a - 2d + 1 + \sqrt{7}$, λ is minimized at either $a = 0$ or $a = \frac{2d-1-\sqrt{7}}{3+\sqrt{7}}$.

In the first case, λ becomes $\frac{3-\sqrt{7}}{2}d(d - 2)$ which is always non-negative for any $d \in \mathbb{Z}_{\geq 0} \setminus \{1\}$.

In the second case, λ becomes $\frac{1}{2}(3(\sqrt{7} - 1)d^2 + 2d(\sqrt{7} - 7) + \sqrt{7} - 5)$ which is always non-negative for any $d \in \mathbb{Z}_{\geq 0} \setminus \{1, 2\}$. For the leftover case $d = 2$, λ becomes $\frac{3-\sqrt{7}}{2}a^2 + (\sqrt{7} - 3)a$ which is always non-negative for any $a \in \mathbb{Z} \setminus \{1\}$. For the other case $d = 1$, λ becomes $\frac{3-\sqrt{7}}{2}a^2 + (\sqrt{7} - 1)a - \frac{3+\sqrt{7}}{2}$ which is always non-negative for any $a \in \mathbb{Z} \setminus \{0, 1\}$. □

Lemma A9. *Let* $\theta : \mathbb{Z}^6_{\geq 0} \to \mathbb{Q}$ *be the function such that* $\theta(a, b, c, d, e, f) = a^2(3 - \sqrt{7}) - a(2d - 1 - \sqrt{7}) + 2bc(\sqrt{7} - 3) + 2(bf + ce) + (d^2 - d)(3 + \sqrt{7}) - 2(3 + \sqrt{7})ef$. *For any* $(a, b, c, d, e, f) \in \mathbb{Z}^6_{\geq 0}$ *such that* $a \geq b + c$ *and* $d \geq e + f$, *it holds that* $\theta(a, b, c, d, e, f) \geq 0$.

Proof. Note first, that for 6-tuples of the form $(0, b, c, 1, e, f)$, it holds that $\theta = 0$, since $a = 0 \Rightarrow b = c = 0$ and $d = 1 \Rightarrow ef = 0$, for 6-tuples of the form $(1, b, c, 1, e, f)$, it holds that $\theta = 2(1 + bf + ce) > 0$, since $a = d = 1 \Rightarrow bc = ef = 0$, and for 6-tuples of the form $(1, b, c, 2, e, f)$, it holds that $\theta = 2bf + 2ce - 2(3 + \sqrt{7})ef + 2(3 + \sqrt{7}) \geq 0$, since $d = 2 \Rightarrow ef \leq 1$. Hence, in the sequel of the proof, we avoid to consider the cases $a = 0 \wedge d = 1$, $a = d = 1$ and $a = 1 \wedge d = 2$.

At a first glance, in order to use standard arguments from calculus, we allow the 6-tuples (a, b, c, d, e, f) to take values in the set of non-negative real numbers. Since it holds that $\frac{\delta\theta}{\delta c} = 2(b(\sqrt{7} - 3) + e)$ and $\frac{\delta\theta}{\delta f} = 2(b - (\sqrt{7} + 3)e)$, θ is minimized at one of the following four cases: $c = 0 \wedge f = 0$, $c = 0 \wedge f = d - e$, $c = a - b \wedge f = 0$ and $c = a - b \wedge f = d - e$.

In the first case, we get $\theta = (3 - \sqrt{7})a^2 + (\sqrt{7} + 1 - 2d)a + (3 + \sqrt{7})(d^2 - d)$. The claim follows by applying Lemma A8, since $\theta \geq \lambda$.

In the second case, we get $\theta = (3 - \sqrt{7})a^2 + (\sqrt{7} + 1 - 2d)a + (3 + \sqrt{7})(d^2 - d) + 2(d - e)(b - (3 + \sqrt{7})e)$. Since $\frac{\delta\theta}{\delta b} = 2(d - e)$, θ is minimized at $b = 0$, which yields $\theta = (3 - \sqrt{7})a^2 + (\sqrt{7} + 1 - 2d)a + (3 + \sqrt{7})(d^2 - d) - 2(d - e)(3 + \sqrt{7})e$. Since $\frac{\delta\theta}{\delta e} = 4(3 + \sqrt{7})e - 2d(3 + \sqrt{7})$, θ is minimized for $e = \frac{d}{2}$. In this case, θ becomes $(3 - \sqrt{7})a^2 + (\sqrt{7} + 1 - 2d)a + (3 + \sqrt{7})(\frac{d^2}{2} - d)$. The claim follows by applying Lemma A8, since $\theta \geq \lambda$.

In the third case, we get $\theta = (3 - \sqrt{7})a^2 + (\sqrt{7} + 1 + 2e - 2d)a + (3 + \sqrt{7})(d^2 - d) + 2b^2(3 - \sqrt{7}) + 2b((\sqrt{7} - 3)a - e)$. Since $\frac{\delta\theta}{\delta e} = 2(a - b)$, θ is minimized for $e = 0$, which yields $\theta = (3 - \sqrt{7})a^2 + (\sqrt{7} + 1 - 2d)a + (3 + \sqrt{7})(d^2 - d) + 2b^2(3 - \sqrt{7}) + 2ab(\sqrt{7} - 3)$. Since $\frac{\delta\theta}{\delta b} = 4(3 - \sqrt{7})b - 2a(3 - \sqrt{7})$, θ is minimized for $b = \frac{a}{2}$. In this case, θ becomes $\frac{3 - \sqrt{7}}{2}a^2 + (\sqrt{7} + 1 - 2d)a + (3 + \sqrt{7})(d^2 - d)$. The claim follows by applying Lemma A8, since $\theta \geq \lambda$.

In the fourth case, we get $\theta = (3 - \sqrt{7})a^2 + (\sqrt{7} + 1 + 2e - 2d)a + (3 + \sqrt{7})(d^2 - d) + 2b^2(3 - \sqrt{7}) + 2(\sqrt{7} - 3)ab + 2bd - 4be + 2(3 + \sqrt{7})e^2$. Since $\frac{\delta\theta}{\delta b} = 4(3 - \sqrt{7})b + 2(\sqrt{7} - 3)a + 2d - 4e$, θ is minimized at either $b = 0$ or $b = \frac{(3 - \sqrt{7})a + 2e - d}{2(3 - \sqrt{7})}$. For $b = 0$, θ becomes $(3 - \sqrt{7})a^2 + (\sqrt{7} + 1 + 2e - 2d)a + (3 + \sqrt{7})(d^2 - d) + 2(3 + \sqrt{7})e^2$. Since $\frac{\delta\theta}{\delta e} = 2a - 2(3 + \sqrt{7})d + 4(3 + \sqrt{7})e$, θ is minimized at either $e = 0$ or $e = \frac{(3 + \sqrt{7})d - a}{2(3 + \sqrt{7})}$. In these two cases, θ becomes, respectively, $(3 - \sqrt{7})a^2 + (\sqrt{7} + 1 - 2d)a + (3 + \sqrt{7})(d^2 - d)$ and $\frac{3}{4}(3 - \sqrt{7})a^2 + (\sqrt{7} + 1 - d)a + (3 + \sqrt{7})(\frac{d^2}{2} - d)$ which are always non-negative because of Lemma A8 and the fact that $\theta \geq \lambda$. For $b = \frac{(3 - \sqrt{7})a + 2e - d}{2(3 - \sqrt{7})}$, θ becomes $\frac{3 - \sqrt{7}}{2}a^2 + (\sqrt{7} + 1 - d)a + (3 + \sqrt{7})(\frac{3d^2}{4} - d) - (3 + \sqrt{7})de + (3 + \sqrt{7})e^2$. Since $\frac{\delta\theta}{\delta e} = (3 + \sqrt{7})(2e - d)$, θ is minimized at either $e = 0$ or $e = \frac{d}{2}$. In these two cases, θ becomes, respectively, $\frac{3 - \sqrt{7}}{2}a^2 + (\sqrt{7} + 1 - d)a + (3 + \sqrt{7})(\frac{3d^2}{4} - d)$ and $\frac{3 - \sqrt{7}}{2}a^2 + (\sqrt{7} + 1 - d)a + (3 + \sqrt{7})(\frac{d^2}{2} - d)$ which are always non-negative because of Lemma A8 and the fact that $\theta \geq \lambda$. $\square$

Lemma A10. *Let $\lambda : \mathbb{Z}_{\geq 0}^2 \to \mathbb{Q}$ be the function such that $\lambda(a, d) = 49a^2 + a(81 - 130d) + 211d^2 - 211d$. For any $(a, d) \in \mathbb{Z}_{\geq 0}^2$, it holds that $\lambda(a, d) \geq 0$.*

Proof. Since $\frac{\delta\lambda}{\delta a} = 98a - 130d + 81$, λ is minimized at either $a = 0$ or $a = \frac{130d - 81}{98}$.

In the first case, λ becomes $d(d - 1)$ which is always non-negative for any $d \in \mathbb{Z}$.

In the second case, λ becomes $d(3057d - 2537) - \frac{6561}{8}$ which is always non-negative for any $d \in \mathbb{Z} \setminus \{0, 1\}$. For the leftover case $d = 0$, λ becomes $49a^2 + 81a$, which is non-negative for any $a \in \mathbb{R}$. For the other case of $d = 1$, λ becomes $a(a - 1)$ which is non-negative for any $a \in \mathbb{Z}$. $\square$

Lemma A11. *Let $\lambda : \mathbb{Z}_{\geq 0}^2 \to \mathbb{Q}$ be the function such that $\lambda(a, d) = 11a^2 + a(81 - 68d) + 422d^2 - 298d$. For any $(a, d) \in \mathbb{Z}_{\geq 0}^2$, it holds that $\lambda(a, d) \geq 0$.*

Proof. Since $\frac{\delta\lambda}{\delta a} = 222a - 2(68d - 81)$, λ is minimized at either $a = 0$ or $a = \frac{68d - 81}{11}$.

In the first case, λ becomes $d(211d - 149)$ which is always non-negative for any $d \in \mathbb{Z}$.

In the second case, λ becomes $d(9d + 3869) - \frac{6561}{2}$ which is always non-negative for any $d \in \mathbb{Z}_{\geq 0} \setminus \{0\}$. For the leftover case of $d = 0$, λ becomes $11a^2 + 162a$ which is non-negative for any $a \in \mathbb{R}$. $\square$

Lemma A12. *Let $\lambda : \mathbb{Z}_{\geq 0}^2 \to \mathbb{Q}$ be the function such that $\lambda(a, d) = 2321a^2 + 422a(81 - 65d) + 84817d^2 - 89042d$. For any $(a, d) \in \mathbb{Z}_{\geq 0}^2 \setminus \{(0, 1)\}$, it holds that $\lambda(a, d) \geq 0$.*

Proof. Since $\frac{\delta\lambda}{\delta a} = 4642a - 422(65d - 81)$, λ is minimized at either $a = 0$ or $a = \frac{65d - 81}{11}$.

In the first case, λ becomes $84817d^2 - 89042d$ which is always non-negative for any $d \in \mathbb{Z} \setminus \{1\}$.

In the second case, λ becomes $d(5189d + 155296) - \frac{1384371}{8}$ which is always non-negative for any $d \in \mathbb{Z}_{\geq 0} \setminus \{0, 1\}$. For the leftover case $d = 0$, λ becomes $11a^2 + 162a$, which is non-negative for any $a \in \mathbb{R}$. For the other case of $d = 1$, λ becomes $a(11a + 32) - \frac{4225}{211}$ which is non-negative for any $a \in \mathbb{Z}_{\geq 0} \setminus \{0\}$. $\quad\square$

Lemma A13. *Let* $\theta : \mathbb{Z}_{\geq 0}^6 \to \mathbb{Q}$ *be the function such that* $\theta(a, b, c, d, e, f) = 49a^2 + a(62b + 62c - 68d - 62e - 62f + 81) + 130bf + 130ce - 422bc + 211d^2 - 149d + 162ef - 62e - 62f$. *For any* $(a, b, c, d, e, f) \in \mathbb{Z}_{\geq 0}^6$ *such that* $a \geq b + c$ *and* $d \geq e + f$, *it holds that* $\theta(a, b, c, d, e, f) \geq 0$.

Proof. At a first glance, in order to use standard arguments from calculus, we allow the 6-tuples (a, b, c, d, e, f) to take values in the set of non-negative real numbers. Since it holds that $\frac{\delta\theta}{\delta c} = 62a - 42b + 130e$ and $\frac{\delta\theta}{\delta f} = -2(31a - 65b - 81e + 31)$, θ is minimized at one of the following four cases: $c = 0 \wedge f = 0$, $c = 0 \wedge f = d - e$, $c = a - b \wedge f = 0$ and $c = a - b \wedge f = d - e$.

In the first case, we get $\theta = 49a^2 + a(62b - 68d - 62e + 81) + 211d^2 - 149d - 62e$. Since $\frac{\delta\theta}{\delta e} = -62(a + 1)$, θ is minimized at $e = d$, which yields $\theta = 49a^2 + a(62b - 130d + 81) + 211d^2 - 211d$. Since $\frac{\delta\theta}{\delta b} = 62a$, θ is minimized for $b = 0$. In this case, θ becomes $49a^2 + a(81 - 130d) + 211d^2 - 211d$. The claim follows by applying Lemma A10.

In the second case, we get $\theta = 49a^2 + a(62b - 130d + 81) + 130b(d - e) + 211d^2 + d(162e - 211) - 162e^2$. Since $\frac{\delta\theta}{\delta b} = 62a + 130(d - e)$, θ is minimized at $b = 0$, which yields $\theta = 49a^2 + a(81 - 130d) + 211d^2 + d(162e - 211) - 162e^2$. Since $\frac{\delta\theta}{\delta e} = 162d - 324e$, θ is minimized at either $e = 0$ and $e = d$. In both cases θ becomes $49a^2 + a(81 - 130d) + 211d^2 - 211d$ and the claim follows by applying Lemma A10.

In the third case, we get $\theta = 111a^2 - a(422b + 68d - 68e - 81) + 422b^2 - 130be + 211d^2 - 149d - 62e$. Since $\frac{\delta\theta}{\delta b} = -422a + 844b - 130e$, θ is minimized at $b = \frac{211a + 65e}{422}$, which yields $\theta = 2321a^2 + 422a(3e + 81 - 68d) + 89042d^2 - 62878d - 4225e^2 - 26164e$. Since $\frac{\delta\theta}{\delta e} = 1266a - 8450e - 26164$, θ is minimized at either $e = 0$ or $e = d$. For $e =$, θ becomes $11a^2 + 2a(81 - 68d) + 422d^2 - 298d$ and the claim follows by applying Lemma A11. For $e = d$, θ becomes $2321a^2 + 422a(81 - 65d) + 84817d^2 - 89042d$ and the claim follows for any 6-tuple (a, b, c, d, e, f) such that $(a, d) \neq (0, 1)$ by applying Lemma A12. Hence, we are left to consider the 6-tuples of the form $(0, 0, 0, 1, e, 0)$. In this case θ becomes $62(1 - e)$ which is always non-negative since $e \in \{0, 1\}$.

In the fourth case, we get $\theta = 111a^2 - a(422b + 130d - 130e - 81) + 422b^2 + 130b(d - 2e) + 211d^2 + d(162e - 211) - 162e^2$. Since $\frac{\delta\theta}{\delta b} = -422a + 844b + 130(d - 2e)$, θ is minimized at either $b = 0$ or $b = \frac{211a - 65(d - 2e)}{422}$. For $b = 0$, θ becomes $111a^2 - a(130d - 130e - 81) + 211d^2 + d(162e - 211) - 162e^2$. Since $\frac{\delta\theta}{\delta e} = 130a + 162d - 324e$, θ is minimized at either $e = 0$ or $e = d$. In these two cases, θ becomes, respectively, $111a^2 + a(81 - 130d) + 211d^2 - 211d$ and $111a^2 + 81a + 211d^2 - 211d$ which are always non-negative because of Lemma A10 and the fact the $\theta \geq \lambda$. For $b = \frac{211a - 65(d - 2e)}{422}$, θ becomes $2321a^2 + 422a(81 - 65d) + 84817d^2 + 2d(42632e - 44521) - 85264e^2$. Since $\frac{\delta\theta}{\delta e} = 85264d - 170528e$, θ is minimized at either $e = 0$ or $e = d$. In both cases, θ becomes $2321a^2 + 422a(81 - 65d) + 84817d^2 - 89042d$ and the claim follows for any 6-tuple (a, b, c, d, e, f) such that $(a, d) \neq (0, 1)$ by applying Lemma A12. Hence, we are left to consider the 6-tuples of the form $(0, 0, 0, 1, e, 1 - e)$. In this case, θ becomes $162(1 - e)$ which is always non-negative since $e \in \{0, 1\}$. $\quad\square$

Appendix B. Missing Proofs

Theorem A1 (Claim of Theorem 4). *There exist two bidimensional linear congestion games* $(\mathcal{G}, \mathcal{C})$ *and* $(\mathcal{G}', \mathcal{C}')$ *such that* $PoA(\mathcal{G}, \mathcal{C}) \geq \frac{119}{33}$ *under the social cost function* Pres *and* $PoA(\mathcal{G}', \mathcal{C}') \geq \frac{35}{8}$ *under the social cost function* Perc.

Proof. For the social cost function Pres, consider the game $(\mathcal{G}, \mathcal{C})$ depicted in Figure A1a). First, we show that σ is a pure Nash equilibrium for $(\mathcal{G}, \mathcal{C})$, that is, no player can lower her perceived cost by switching to her optimal strategy. Player 1 is paying $27 \cdot 2 + 46 = 100$; by switching to σ_1^*, she pays $7 \cdot 4 + 18 \cdot 4 = 100$. Player 2 is paying $27 \cdot 2 + 42 + 56 = 152$; by switching to σ_2^*, she pays $17 \cdot 4 + 21 \cdot 4 = 152$. Player 3 is paying $27 \cdot 2 + 42 = 96$; by switching to σ_3^*, she pays $7 \cdot 4 + 17 \cdot 4 = 96$. Player 4 is paying $27 \cdot 2 + 46 + 56 = 156$; by switching to σ_4^*, she pays $18 \cdot 4 + 21 \cdot 4 = 156$.

Player 5 is paying $7 \cdot 3 + 17 \cdot 3 + 21 \cdot 3 = 135$; by switching to σ_5^*, she pays $27 \cdot 5 = 135$. Player 6 is paying $7 \cdot 3 + 18 \cdot 3 + 21 \cdot 3 = 138$; by switching to σ_6^*, she pays $46 \cdot 3 = 138$. Player 7 is paying $7 \cdot 3 + 18 \cdot 3 + 17 \cdot 3 = 126$; by switching to σ_7^*, she pays $42 \cdot 3 = 126$. Player 8 is paying $18 \cdot 3 + 17 \cdot 3 + 21 \cdot 3 = 168$; by switching to σ_8^*, she pays $56 \cdot 3 = 168$.

The price of anarchy of $(\mathcal{G}, \mathcal{C})$ is then lower bounded by the ratio

$$\frac{100 + 152 + 96 + 156 + 135 + 138 + 126 + 168}{25 + 38 + 24 + 39 + 27 + 46 + 42 + 56} = \frac{1071}{297} = \frac{119}{33}.$$

For the social cost function Perc, consider the game $(\mathcal{G}', \mathcal{C}')$ depicted in Figure A1b). First, we show that σ is a pure Nash equilibrium for $(\mathcal{G}', \mathcal{C}')$, that is, no player can lower her perceived cost by switching to her optimal strategy. Player 1 is paying $1418 + 958 + 189 \cdot 2 = 2754$; by switching to σ_1^*, she pays $918 \cdot 3 = 2754$. Player 2 is paying $616 + 221 + 189 \cdot 2 = 1215$; by switching to σ_2^*, she pays $405 \cdot 3 = 1215$. Player 3 is paying $1418 + 616 + 189 \cdot 2 = 2412$; by switching to σ_3^*, she pays $804 \cdot 3 = 2412$. Player 4 is paying $958 + 221 + 189 \cdot 2 = 1557$; by switching to σ_4^*, she pays $519 \cdot 3 = 1557$. Player 5 is paying $(918 + 405 + 804) \cdot 2 = 4254$; by switching to σ_5^*, she pays $1418 \cdot 3 = 4254$. Player 6 is paying $(918 + 519) \cdot 2 = 2874$; by switching to σ_6^*, she pays $958 \cdot 3 = 2874$. Player 7 is paying $(405 + 519) \cdot 2 = 1848$; by switching to σ_7^*, she pays $616 \cdot 3 = 1848$. Player 8 is paying $804 \cdot 2 = 1608$; by switching to σ_8^*, she pays $221 \cdot 3 + 189 \cdot 5 = 1608$.

By noting that the perceived cost of the first four players is exactly twice their presumed one, the price of anarchy of $(\mathcal{G}', \mathcal{C}')$ is then lower bounded by the ratio

$$\frac{2 \cdot (2754 + 1215 + 2412 + 1557) + 4254 + 2874 + 1848 + 1608}{1418 + 958 + 616 + 221 + 189 + 918 + 405 + 804 + 519} = \frac{26460}{6048} = \frac{35}{8}.$$

$\square$

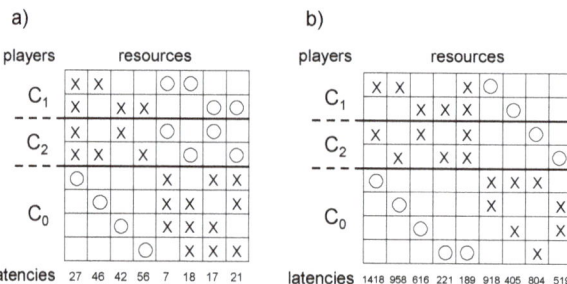

a)   b)

Figure A1. The games depicted in figures (**a**,**b**) represent the lower bound instances w.r.t. the social cost functions Pres and Perc, respectively. Each column in the matrix represents a resource of cost function $\ell(x) = \alpha x$ whose coefficient α is reported at the bottom of the column. Each row i in the matrix models the strategy set of player i as follows: circles represent resources belonging to σ_i, while crosses represent resources belonging to σ_i^*.

Theorem A2 (Claim of Theorem 6). *For any $\epsilon > 0$, there exist two bidimensional linear congestion games $(\mathcal{G}, \mathcal{C})$ and $(\mathcal{G}', \mathcal{C}')$ such that $PoS(\mathcal{G}, \mathcal{C}) \geq \frac{1+\sqrt{5}}{2} - \epsilon$ under the social cost function Pres and $PoS(\mathcal{G}', \mathcal{C}') \geq \frac{5+\sqrt{17}}{4} - \epsilon$ under the social cost function Perc.*

Proof. Let $(\mathcal{G}, \mathcal{C})$ be a bidimensional linear congestion game such that $|C_0| = n_0$ and $|C_1| = |C_2| = n_1$. Each player $i \in C_1 \cup C_2$ has two strategies σ_i and σ_i^*, while all players in C_0 have the same strategy $\bar{s}$. There are three types of resources:

- n_1 resources r_i, $i \in [n_1]$, each with cost function $\ell_{r_i}(x) = \frac{n_1 + 2n_0 + 1 + \gamma}{2} x$, where γ is an arbitrarily small positive value. Resource r_i belongs only to σ_i^*;
- $n_1(n_1 - 1)$ resources $r'_{i,j}$, $i, j \in [n_1]$ with $i \neq j$, each with cost function $\ell_{r'_{ij}}(x) = \frac{1}{2}x$. Resource r'_{ij} belongs only to σ_i and to σ_j^*;
- one resource r'' with cost function $\ell_{r''}(x) = 1$. Resource r'' belongs to σ_i for each $i \in [n_1]$ and to $\bar{s}$.

Let σ (resp. σ^*) be the strategy profile in which each player $i \notin C_0$ plays strategy σ_i (resp. σ_i^*). The cost of each player $i \in C_j$, with $j \in \{1, 2\}$, for adopting strategy σ_i when there are exactly h players in C_j adopting the strategy played in σ (and thus there are $n_1 - h$ players in C_j adopting the strategy played in σ^*) is $cost_\sigma(h) = \frac{2n_1 - h - 1}{2} + n_0 + h$. Similarly, the cost of each player $i \in C_j$ for adopting strategy σ_i^* when there are exactly h players in C_j adopting the strategy played in σ is $cost_{\sigma^*}(h) = \frac{n_1 + 2n_0 + 1 + \gamma}{2} + \frac{n_1 + h - 1}{2}$. Since for any $h \in [n_1]$, it holds that $cost_{\sigma^*}(h - 1) > cost_\sigma(h)$, it follows that σ is the only pure Nash equilibrium for $(\mathcal{G}, \mathcal{C})$.

The price of stability of $(\mathcal{G}, \mathcal{C})$ is then lower bounded by the following ratio

$$\frac{n_1(n_1 - 1) + 2n_1(n_1 + n_0) + n_0(2n_1 + n_0)}{n_1(n_1 + 2n_0 + 1 + \gamma) + n_1(n_1 - 1) + n_0^2},$$

which, for n_0 going to infinity and $n_1 = \frac{1 + \sqrt{5}}{2} n_0$, tends to $\frac{1 + \sqrt{5}}{2}$.

Let $(\mathcal{G}', \mathcal{C}')$ be a bidimensional linear congestion game such that $C_0 = \varnothing$, $|C_1| = n_1$ and $|C_2| = n_2$. Each player $i \in C_1$ has two strategies σ_i and σ_i^*, while all players in C_2 have the same strategy $\bar{s}$.

There are three types of resources:

- n_1 resources r_i, $i \in [n_1]$, each with cost function $\ell_{r_i}(x) = \frac{n_1 + 1 + \gamma}{2} x$, where γ is an arbitrarily small positive value. Resource r_i belongs only to σ_i^*;
- $n_1(n_1 - 1)$ resources $r'_{i,j}$, $i, j \in [n_1]$ with $i \neq j$, each with cost function $\ell_{r'_{ij}}(x) = \frac{1}{2}x$. Resource r'_{ij} belongs only to σ_i and to σ_j^*;
- one resource r'' with cost function $\ell_{r''}(x) = 1$. Resource r'' belongs to σ_i for each $i \in [n_1]$ and to $\bar{s}$.

Let σ (resp. σ^*) be the strategy profile in which each player $i \notin C_0$ plays strategy σ_i (resp. σ_i^*). The cost of each player $i \in C_1$ for adopting strategy σ_i when there are exactly h players in C_1 adopting the strategy played in σ (and thus there are $n_1 - h$ players in C_1 adopting the strategy played in σ^*) is $cost_\sigma(h) = \frac{2n_1 - h - 1}{2} + h$. Similarly, the cost of each player $i \in C_1$ for adopting strategy σ_i^* when there are exactly h players in C_1 adopting the strategy played in σ is $cost_{\sigma^*}(h) = \frac{n_1 + 1 + \gamma}{2} + \frac{n_1 + h - 1}{2}$. Since for any $h \in [n_1]$, it holds that $cost_{\sigma^*}(h - 1) > cost_\sigma(h)$, it follows that σ is the only pure Nash equilibrium for $(\mathcal{G}', \mathcal{C}')$.

The price of stability of $(\mathcal{G}', \mathcal{C}')$ is then lower bounded by the following ratio

$$\frac{\frac{1}{2}n_1(n_1 - 1) + (n_1 + n_2)^2}{\frac{1}{2}n_1(n_1 + 1 + \gamma) + \frac{1}{2}n_1(n_1 - 1) + n_2^2},$$

which, for n_2 going to infinity and $n_1 = \frac{1 + \sqrt{17}}{4} n_2$, tends to $\frac{1 + \sqrt{17}}{4}$. $\quad\square$

References

1. Beckmann, M.J.; McGuire, C.B.; Winsten, C.B. *Studies in the Economics of Transportation*; Yale University Press: London, UK, 1956.
2. Rosenthal, R.W. A Class of Games Possessing Pure-Strategy Nash Equilibria. *Int. J. Game Theory* **1973**, *2*, 65–67. [CrossRef]
3. Wardrop, J.G. Road Paper. Some Theoretical Aspects of Road Traffic Research. *Proc. Inst. Civ. Eng.* **1952**, *1*, 325–362. [CrossRef]
4. Wardrop, J.G.; Whitehead, J.I. Correspondence. Some Theoretical Aspects of Road Traffic Research. *Proc. Inst. Civ. Eng.* **1952**, *1*, 767–768. [CrossRef]

5. Nash, J.F. Equilibrium points in *n*-person games. *Proc. Natl. Acad. Sci. USA* **1950**, *36*, 48–49. [CrossRef] [PubMed]

6. Koutsoupias, E.; Papadimitriou, C. Worst-case equilibria. In Proceedings of the 16th International Symposium on Theoretical Aspects of Computer Science (STACS), LNCS 1653, Trier, Germany, 4–6 March 1999; pp. 404–413.

7. Anshelevich, E.; Dasgupta, A.; Kleinberg, J.; Tardos, E.; Wexler, T.; Roughgarden, T. The Price of Stability for Network Design with Fair Cost Allocation. *SIAM J. Comput.* **2008**, *38*, 1602–1623. [CrossRef]

8. Aland, S.; Dumrauf, D.; Gairing, M.; Monien, B.; Schoppmann, F. Exact Price of Anarchy for Polynomial Congestion Games. *SIAM J. Comput.* **2011**, *40*, 1211–1233. [CrossRef]

9. Awerbuch, B.; Azar, Y.; Epstein, A. The Price of Routing Unsplittable Flow. *SIAM J. Comput.* **2013**, *42*, 160–177. [CrossRef]

10. Bhawalkar, K.; Gairing, M.; Roughgarden, T. Weighted Congestion Games: Price of Anarchy, Universal Worst-Case Examples, and Tightness. *ACM Trans. Econ. Comput.* **2014**, *2*, 14:1–14:23. [CrossRef]

11. Bilò, V. A Unifying Tool for Bounding the Quality of Non-Cooperative Solutions in Weighted Congestion Games. *Theory Comput. Syst.* **2018**, *62*, 1288–1317. [CrossRef]

12. Caragiannis, I.; Flammini, M.; Kaklamanis, C.; Kanellopoulos, P.; Moscardelli, L. Tight Bounds for Selfish and Greedy Load Balancing. *Algorithmica* **2011**, *61*, 606–637. [CrossRef]

13. Christodoulou, G.; Gairing, M. Price of Stability in Polynomial Congestion Games. *ACM Trans. Econ. Comput.* **2016**, *4*, 10:1–10:17. [CrossRef]

14. Christodoulou, G.; Gairing, M.; Giannakopoulos, Y.; Spirakis, P.G. The Price of Stability of Weighted Congestion Games. *SIAM J. Comput.* **2019**, *48*, 1544–1582. [CrossRef]

15. Christodoulou, G.; Koutsoupias, E. The Price of Anarchy of Finite Congestion Games. In Proceedings of the 37th Annual ACM Symposium on Theory of Computing (STOC), Baltimore, MD, USA, 22–24 May 2005; pp. 67–73,

16. Christodoulou, G.; Koutsoupias, E. On the Price of Anarchy and Stability of Correlated Equilibria of Linear Congestion Games. In Proceedings of the 13th Annual European Symposium on Algorithms (ESA), LNCS 3669, Palma de Mallorca, Spain, 3–6 October 2005; pp. 59–70.

17. Christodoulou, G.; Koutsoupias, E.; Spirakis, P.G. On the Performance of Approximate Equilibria in Congestion Games. *Algorithmica* **2011**, *61*, 116–140. [CrossRef]

18. Bilò, V.; Vinci, C. On the Impact of Singleton Strategies in Congestion Games. In Proceedings of the 25th Annual European Symposium on Algorithms (ESA), LIPIcs, Vienna, Austria, 4–6 September 2017; pp. 17:1–17:14.

19. Roughgarden, T. Intrinsic Robustness of the Price of Anarchy. *J. ACM* **2015**, *62*, 32:1–32:42. [CrossRef]

20. Fotakis, D.; Kontogiannis, S.; Spirakis, P. Selfish Unsplittable Flows. *Theor. Comput. Sci.* **2005**, *348*, 226–239. [CrossRef]

21. Harks, T.; Klimm, M. On the Existence of Pure Nash Equilibria in Weighted Congestion Games. *Math. Oper. Res.* **2012**, *37*, 419–436. [CrossRef]

22. Harks, T.; Klim, M.; Möhring, R.H. Characterizing the Existence of Potential Functions in Weighted Congestion Games. *Theory Comput. Syst.* **2011**, *49*, 46–70. [CrossRef]

23. Bilò, V.; Fanelli, A.; Flammini, M.; Moscardelli, L. When Ignorance Helps: Graphical Multicast Cost Sharing Games. *Theor. Comput. Sci.* **2010**, *411*, 660–671. [CrossRef]

24. Bilò, V.; Fanelli, A.; Flammini, M.; Moscardelli, L. Graphical Congestion Games. *Algorithmica* **2011**, *61*, 274–297. [CrossRef]

25. Fotakis, D.; Gkatzelis, V.; Kaporis, A.C.; Spirakis, P.G. The Impact of Social Ignorance on Weighted Congestion Games. *Theory Comput. Syst.* **2012**, *50*, 559–578. [CrossRef]

26. Bilò, V.; Flammini, M.; Gallotti, V. On Bidimensional Congestion Games. In *Proceedings of the 19th International Colloquium on Structural Information and Communication Complexity (SIROCCO)*, LNCS 7355, Reykjavik, Iceland, 30 June–2 July 2012; pp. 147–158.

27. Bilò, V.; Vinci, C. Dynamic Taxes for Polynomial Congestion Games. *ACM Trans. Econ. Comput.* **2019**, *7*, 15:1–15:36. [CrossRef]

28. Caragiannis, I.; Kaklamanis, C.; Kanellopoulos, P. Taxes for linear atomic congestion games. *ACM Trans. Algorithms* **2010**, *7*, 13:1–13:31. [CrossRef]

29. Bilò, V.; Vinci, C. On Stackelberg Strategies in Affine Congestion Games. *Theory Comput. Syst.* **2019**, *63*, 1228–1249. [CrossRef]

30. Fotakis, D. Stackelberg Strategies for Atomic Congestion Games. *Theory Comput. Syst.* **2010**, *47*, 218–249. [CrossRef]

31. Bell, M.G. Hyperstar: A Multi-path Astar Algorithm for Risk Averse Vehicle Navigation. *Transp. Res. Part B Methodol.* **2009**, *43*, 97–107. [CrossRef]

32. Bell, M.G.; Cassir, C. Risk-averse User Equilibrium Traffic Assignment: An Application of Game Theory. *Transp. Res. Part B Methodol.* **2002**, *36*, 671–681. [CrossRef]

33. Yekkehkhany, A.; Murray, T.; Nagi, R. Road Paper. Risk-Averse Equilibrium for Games. *arXiv* **2020**, arXiv:2002.08414.

34. Yekkehkhany, A.; Nagi, R. Risk-Averse Equilibrium for Autonomous Vehicles in Stochastic Congestion Games. *arXiv* **2020**, arXiv:2007.09771.

35. Bilò, V.; Moscardelli, L.; Vinci, C. Uniform Mixed Equilibria in Network Congestion Games with Link Failures. In Proceedings of the 45th International Colloquium on Automata, Languages, and Programming (ICALP), LIPIcs 107, Prague, Czech Republic, 9–13 July 2018; pp. 146:1–146:14.

36. Penn, M.; Polukarov, M.; Tennenholtz, M. Congestion Games with Failures. *Discret. Appl. Math.* **2011**, *159*, 1508–1525. [CrossRef]

37. Penn, M.; Polukarov, M.; Tennenholtz, M. Congestion Games with Load-dependent Failures: Identical Resources. *Games Econ. Behav.* **2009**, *67*, 156–173. [CrossRef]

Publisher's Note: MDPI stays neutral with regard to jurisdictional claims in published maps and institutional affiliations.

 **algorithms**

 MDPI

Article

Algorithmic Aspects of Some Variations of Clique Transversal and Clique Independent Sets on Graphs

Chuan-Min Lee

Department of Computer and Communication Engineering, Ming Chuan University, 5 De Ming Road, Guishan District, Taoyuan City 333, Taiwan; joneslee@mail.mcu.edu.tw; Tel.: +886-3-350-7001 (ext. 3432); Fax: +886-3-359-3876

Abstract: This paper studies the maximum-clique independence problem and some variations of the clique transversal problem such as the $\{k\}$-clique, maximum-clique, minus clique, signed clique, and k-fold clique transversal problems from algorithmic aspects for k-trees, suns, planar graphs, doubly chordal graphs, clique perfect graphs, total graphs, split graphs, line graphs, and dually chordal graphs. We give equations to compute the $\{k\}$-clique, minus clique, signed clique, and k-fold clique transversal numbers for suns, and show that the $\{k\}$-clique transversal problem is polynomial-time solvable for graphs whose clique transversal numbers equal their clique independence numbers. We also show the relationship between the signed and generalization clique problems and present NP-completeness results for the considered problems on k-trees with unbounded k, planar graphs, doubly chordal graphs, total graphs, split graphs, line graphs, and dually chordal graphs.

Keywords: clique independent set; clique transversal number; signed clique transversal function; minus clique transversal function; k-fold clique transversal set

 check for updates

Citation: Lee, C.-M. Algorithmic Aspects of Some Variations of Clique Transversal and Clique Independent Sets on Graphs. *Algorithms* **2021**, *14*, 22. https://doi.org/10.3390/a14010022

Received: 9 December 2020
Accepted: 11 January 2021
Published: 13 January 2021

Publisher's Note: MDPI stays neutral with regard to jurisdictional claims in published maps and institutional affiliations.

1. Introduction

Every graph $G = (V, E)$ in this paper is finite, undirected, connected, and has at most one edge between any two vertices in G. We assume that the vertex set V and edge set E of G contain n vertices and m edges. They can also be denoted by $V(G)$ and $E(G)$. A graph $G' = (V', E')$ is an *induced subgraph* of G denoted by $G[V']$ if $V' \subseteq V$ and E' contains all the edge $(x, y) \in E$ for $x, y \in V'$. Two vertices $x, y \in V$ are *adjacent* or *neighbors* if $(x, y) \in E$. The sets $N_G(x) = \{y \mid (x, y) \in E\}$ and $N_G[x] = N_G(x) \cup \{x\}$ are the *neighborhood* and *closed neighborhood* of a vertex x in G, respectively. The number $deg_G(x) = |N_G(x)|$ is the *degree* of x in G. If $deg_G(x) = k$ for every $x \in V$, then G is k-*regular*. Particularly, *cubic graphs* are an alternative name for 3-regular graphs.

A subset S of V is a *clique* if $(x, y) \in E$ for $x, y \in S$. Let Q be a clique of G. If $Q \cap Q' \neq Q$ for any other clique Q' of G, then Q is a *maximal* clique. We use $C(G)$ to represent the set $\{C \mid C$ is a maximal clique of $G\}$. A clique $S \in C(G)$ is a *maximum* clique if $|S| \geq |S'|$ for every $S' \in C(G)$. The number $\omega(G) = \max\{|S| \mid S \in C(G)\}$ is the *clique number* of G. A set $D \subseteq V$ is a *clique transversal set* (abbreviated as CTS) of G if $|C \cap D| \geq 1$ for every $C \in C(G)$. The number $\tau_C(G) = \min\{|S| \mid S$ is a CTS of $G\}$ is the *clique transversal number* of G. The *clique transversal problem* (abbreviated as CTP) is to find a minimum CTS for a graph. A set $S \subseteq C(G)$ is a *clique independent set* (abbreviated as CIS) of G if $|S| = 1$ or $|S| \geq 2$ and $C \cap C' = \emptyset$ for $C, C' \in S$. The number $\alpha_C(G) = \max\{|S| \mid S$ is a CIS of $G\}$ is the *clique independence number* of G. The *clique independence problem* (abbreviated as CIP) is to find a maximum CIS for a graph.

The CTP and the CIP have been widely studied. Some studies on the CTP and the CIP consider imposing some additional constraints on CTS or CIS, such as the *maximum-clique independence problem* (abbreviated as MCIP), the k-*fold clique transversal problem* (abbreviated as k-FCTP), and the *maximum-clique transversal problem* (abbreviated as MCTP).

Definition 1 ([1,2]). *Suppose that $k \in \mathbb{N}$ is fixed and G is a graph. A set $D \subseteq V(G)$ is a k-fold clique transversal set (abbreviated as k-FCTS) of G if $|C \cap D| \geq k$ for $C \in C(G)$. The number $\tau_C^k(G) = min\{|S| \mid S \text{ is a } k\text{-FCTS of } G\}$ is the k-fold clique transversal number of G. The k-FCTP is to find a minimum k-FCTS for a graph.*

Definition 2 ([3,4]). *Suppose that G is a graph. A set $D \subseteq V(G)$ is a maximum-clique transversal set (abbreviated as MCTS) of G if $|C \cap D| \geq 1$ for $C \in C(G)$ with $|C| = \omega(G)$. The number $\tau_M(G) = min\{|S| \mid S \text{ is an MCTS of } G\}$ is the maximum-clique transversal number of G. The MCTP is to find a minimum MCTS for a graph. A set $S \subseteq C(G)$ is a maximum-clique independent set (abbreviated as MCIS) of G if $|C| = \omega(G)$ for $C \in S$ and $C \cap C' = \emptyset$ for $C, C' \in S$. The number $\alpha_M(G) = max\{|S| \mid S \text{ is an MCIS of } G\}$ is the maximum-clique independence number of G. The MCIP is to find a maximum MCIS for a graph.*

The k-FCTP on balanced graphs can be solved in polynomial time [2]. The MCTP has been studied in [3] for several well-known graph classes and the MCIP is polynomial-time solvable for any graph H with $\tau_M(H) = \alpha_M(H)$ [4]. Assume that $Y \subseteq \mathbb{R}$ and $f : X \to Y$ is a function. Let $f(X') = \sum_{x \in X} f(x)$ for $X' \subseteq X$, and let $f(X)$ be the *weight* of f. A CTS of G can be expressed as a function f whose domain is $V(G)$ and range is $\{0, 1\}$, and $f(C) \geq 1$ for $C \in C(G)$. Then, f is a *clique transversal function* (abbreviated as CTF) of G and $\tau_C(G) = min\{f(V(G)) \mid f \text{ is a CTF of } G\}$. Several types of CTF have been studied [4–7]. The following are examples of CTFs.

Definition 3. *Suppose that $k \in \mathbb{N}$ is fixed and G is a graph. A function f is a $\{k\}$-clique transversal function (abbreviated as $\{k\}$-CTF) of G if the domain and range of f are $V(G)$ and $\{0, 1, 2, \ldots, k\}$, respectively, and $f(C) \geq k$ for $C \in C(G)$. The number $\tau_C^{\{k\}}(G) = min\{f(V(G)) \mid f \text{ is a } \{k\}\text{-CTF of } G\}$ is the $\{k\}$-clique transversal number of G. The $\{k\}$-clique transversal problem (abbreviated as $\{k\}$-CTP) is to find a minimum-weight $\{k\}$-CTF for a graph.*

Definition 4. *Suppose that G is a graph. A function f is a signed clique transversal function (abbreviated as SCTF) of G if the domain and range of f are $V(G)$ and $\{-1, 1\}$, respectively, and $f(C) \geq 1$ for $C \in C(G)$. If the domain and range of f are $V(G)$ and $\{-1, 0, 1\}$, respectively, and $f(C) \geq 1$ for $C \in C(G)$, then f is a minus clique transversal function (abbreviated as MCTF) of G. The number $\tau_C^s(G) = min\{f(V(G)) \mid f \text{ is an SCTF of } G\}$ is the signed clique transversal number of G. The minus clique transversal number of G is $\tau_C^-(G) = min\{f(V(G)) \mid f \text{ is an MCTF of } G\}$. The signed clique transversal problem (abbreviated as SCTP) is to find a minimum-weight SCTF for a graph. The minus clique transversal problem (abbreviated as MCTP) is to find a minimum-weight MCTF for a graph.*

Lee [4] introduced some variations of the k-FCTP, the $\{k\}$-CTP, the SCTP, and the MCTP, but those variations are dedicated to maximum cliques in a graph. The MCTP on chordal graphs is NP-complete, while the MCTP on block graphs is linear-time solvable [7]. The MCTP and SCTP are linear-time solvable for any strongly chordal graph G if a *strong elimination ordering* of G is given [5]. The SCTP is NP-complete for doubly chordal graphs [6] and planar graphs [5].

According to what we have described above, there are very few algorithmic results regarding the k-FCTP, the $\{k\}$-CTP, the SCTP, and the MCTP on graphs. This motivates us to study the complexities of the k-FCTP, the $\{k\}$-CTP, the SCTP, and the MCTP. This paper also studies the MCTP and MCIP for some graphs and investigates the relationships between different *dominating functions* and CTFs.

Definition 5. *Suppose that $k \in \mathbb{N}$ is fixed and G is a graph. A set $S \subseteq V(G)$ is a k-tuple dominating set (abbreviated as k-TDS) of G if $|S \cap N_G[x]| \geq 1$ for $x \in V(G)$. The number $\gamma_{\times k}(G) = min\{|S| \mid S \text{ is a } k\text{-TDS of } G\}$ is the k-tuple domination number of G. The k-tuple domination problem (abbreviated as k-TDP) is to find a minimum k-TDS for a graph.*

Notice that a *dominating set* of a graph G is a 1-TDS. The *domination number* $\gamma(G)$ of G is $\gamma_{\times 1}(G)$.

Definition 6. *Suppose that $k \in \mathbb{N}$ is fixed and G is a graph. A function f is a $\{k\}$-dominating function (abbreviated as $\{k\}$-DF) of G if the domain and range of f are $V(G)$ and $\{0, 1, 2, \ldots, k\}$, respectively, and $f(N_G[x]) \geq k$ for $x \in V(G)$. The number $\gamma_{\{k\}}(G) = \min\{f(V(G)) \mid f \text{ is a } \{k\}\text{-DF of } G\}$ is the $\{k\}$-domination number of G. The $\{k\}$-domination problem (abbreviated as $\{k\}$-DP) is to find a minimum-weight $\{k\}$-DF for a graph.*

Definition 7. *Suppose that G is a graph. A function f is a signed dominating function (abbreviated as SDF) of G if the domain and range of f are $V(G)$ and $\{-1, 1\}$, respectively, and $f(N_G[x]) \geq 1$ for $x \in V(G)$. If the domain and range of f are $V(G)$ and $\{-1, 0, 1\}$, respectively, and $f(N_G[x]) \geq 1$ for $x \in V(G)$, then f is a minus dominating function (abbreviated as MDF) of G. The number $\gamma_s(G) = \min\{f(V(G)) \mid f \text{ is an SDF of } G\}$ is the signed domination number of G. The minus domination number of G is $\gamma^-(G) = \min\{f(V(G)) \mid f \text{ is an MDF of } G\}$. The signed domination problem (abbreviated as SDP) is to find a minimum-weight SDF for a graph. The minus domination problem (abbreviated as MDP) is to find a minimum-weight MDF for a graph.*

Our main contributions are as follows.

1. We prove in Section 2 that $\gamma^-(G) = \tau_C^-(G)$ and $\gamma_s(G) = \tau_C^s(G)$ for any sun G. We also prove that $\gamma_{\times k}(G) = \tau_C^k(G)$ and $\gamma_{\{k\}}(G) = \tau_C^{\{k\}}(G)$ for any sun G if $k > 1$.

2. We prove in Section 3 that $\tau_C^{\{k\}}(G) = k\tau_C(G)$ for any graph G with $\tau_C(G) = \alpha_C(G)$. Then, $\tau_C^{\{k\}}(G)$ is polynomial-time solvable if $\tau_C(G)$ can be computed in polynomial time. We also prove that the SCTP is a special case of *the generalized clique transversal problem* [8]. Therefore, the SCTP for a graph H can be solved in polynomial time if the generalized transversal problem for H is polynomial-time solvable.

3. We show in Section 4 that $\gamma_{\times k}(G) = \tau_C^k(G)$ and $\gamma_{\{k\}}(G) = \tau_C^{\{k\}}(G)$ for any split graph G. Furthermore, we introduce H_1-*split graphs* and prove that $\gamma^-(H) = \tau_C^-(H)$ and $\gamma_s(H) = \tau_C^s(H)$ for any H_1-split graph H. We prove the NP-completeness of SCTP for split graphs by showing that the SDP on H_1-split graphs is NP-complete.

4. We show in Section 5 that $\tau_C^{\{k\}}(G)$ for a *doubly chordal graph* G can be computed in linear time, but the k-FCTP is NP-complete for doubly chordal graphs as $k > 1$. Notice that the CTP is a special case of the k-FCTP and the $\{k\}$-CTP when $k = 1$, and thus $\tau_C(G) = \tau_C^1(G) = \tau_C^{\{1\}}(G)$ for any graph G.

5. We present other NP-completeness results in Sections 6 and 7 for k-trees with unbounded k and subclasses of total graphs, line graphs, and planar graphs. These results can refine the "borderline" between P and NP for the considered problems and graphs classes or their subclasses.

2. Suns

In this section, we give equations to compute $\tau_C^{\{k\}}(G)$, $\tau_C^k(G)$, $\tau_C^s(G)$, and $\tau_C^-(G)$ for any sun G and show that $\tau_C^{\{k\}}(G) = \gamma_{\{k\}}(G)$, $\tau_C^k(G) = \gamma_{\times k}(G)$, $\tau_C^s(G) = \gamma_s(G)$, and $\tau_C^-(G) = \gamma^-(G)$.

Let $p \in \mathbb{N}$ and G be a graph. An edge $e \in E(G)$ is a *chord* if e connects two non-consecutive vertices of a cycle in G. If C has a chord for every cycle C consisting of more than three vertices, G is a *chordal* graph. A *sun* G is a chordal graph whose vertices can be partitioned into $W = \{w_i \mid 1 \leq i \leq p\}$ and $U = \{u_i \mid 1 \leq i \leq p\}$ such that (1) W is an independent set, (2) the vertices $u_1, u_2, \ldots, u_p$ of U form a cycle, and (3) every $w_i \in W$ is adjacent to precisely two vertices u_i and u_j, where $j \equiv i + 1 \pmod p$. We use $S_p = (W, U, E)$ to denote a sun. Then, $|V(S_p)| = 2p$. If p is odd, S_p is an *odd* sun; otherwise, it is an *even* sun. Figure 1 shows two suns.

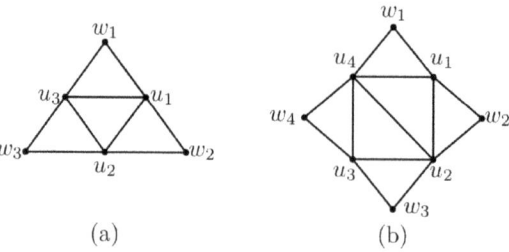

Figure 1. (a) The sun S_3. (b) A sun S_4.

Lemma 1. *For any sun $S_p = (W, U, E)$, $\tau_C^2(S_p) = p$ and $\tau_C^3(S_p) = 2p$.*

Proof. It is straightforward to see that U is a minimum 2-FCTS and $W \cup U$ is a minimum 3-FCTS of S_p. This lemma therefore holds. □

Lemma 2. *Suppose that $k \in \mathbb{N}$ and $k > 1$. Then, $\tau_C^{\{k\}}(S_p) = \lceil pk/2 \rceil$ for any sun $S_p = (W, U, E)$.*

Proof. Let $i, j \in \{1, 2, \dots p\}$ such that $j \equiv i + 1 \pmod{p}$. Since every $w_i \in W$ is adjacent to precisely two vertices $u_i, u_j \in U$, $N_{S_p}[w_i] = \{w_i, u_i, u_j\}$ is a maximal clique of S_p. By contradiction, we can prove that there exists a minimum $\{k\}$-CTF f of S_p such that $f(w_i) = 0$ for $w_i \in W$. Since $f(N_{S_p}[w_i]) \geq k$ for $1 \leq i \leq p$, we have

$$\tau_C^{\{k\}}(S_p) = \sum_{i=1}^{p} f(u_i) = \frac{\sum_{i=1}^{p} f(N_{S_p}[w_i])}{2} \geq \frac{pk}{2}.$$

Since $\tau_C^{\{k\}}(S_p)$ is a nonnegative integer, $\tau_C^{\{k\}}(S_p) \geq \lceil pk/2 \rceil$.

We define a function $h : W \cup U \to \{0, 1, \dots, k\}$ by $h(w_i) = 0$ for every $w_i \in W$, $h(u_i) = \lceil k/2 \rceil$ for $u_i \in U$ with odd index i and $h(u_i) = \lfloor k/2 \rfloor$ for every $u_i \in U$ with even index i. Clearly, a maximal clique Q of S_n is either the closed neighborhood of some vertex in W or a set of at least three vertices in U. If $Q = N_{S_p}[w_i]$ for some $w_i \in W$, then $h(Q) = \lceil k/2 \rceil + \lfloor k/2 \rfloor = k$. Suppose that Q is a set of at least three vertices in U. Since $k \geq 2$, $h(Q) \geq 3 \cdot \lfloor k/2 \rfloor \geq k$. Therefore, h is a $\{k\}$-CTF of S_p. We show the weight of h is $\lceil pk/2 \rceil$ by considering two cases as follows.

Case 1: p is even. We have

$$h(V(S_p)) = \sum_{i=1}^{p} h(u_i) = \frac{p}{2} \cdot (\lceil k/2 \rceil + \lfloor k/2 \rfloor) = \frac{pk}{2}.$$

Case 2: p is odd. We have

$$h(V(S_p)) = \sum_{i=1}^{p} h(u_i) = \frac{(p-1)}{2} \cdot k + \lceil k/2 \rceil = \lceil pk/2 \rceil.$$

Following what we have discussed above, we know that h is a minimum $\{k\}$-CTF of S_n and thus $\tau_C^{\{k\}}(S_p) = \lceil pk/2 \rceil$. □

Lemma 3. *For any sun $S_p = (W, U, E)$, $\tau_C^-(S_p) = \tau_C^s(S_p) = 0$.*

Proof. For $1 \le i \le p$, $N_{S_p}[w_i]$ is a maximal clique of S_p. Let h be a minimum SCTF of S_p. Then, $\tau_C^s(S_p) = h(V(S_p))$. Note that $h(N_{S_p}[w_i]) \ge 1$ for $1 \le i \le p$. We have

$$h(V(S_p)) = \sum_{i=1}^{p} h(N_{S_p}[w_i]) - \sum_{i=1}^{p} h(u_i) \ge p - \sum_{i=1}^{p} h(u_i).$$

Since $\sum_{i=1}^{p} h(u_i) \le p$, we have $\tau_C^s(S_p) \ge 0$. Let f be an SCTF of S_p such that $f(u_i) = 1$ and $f(w_i) = -1$ for $1 \le i \le p$. The weight of f is 0. Then f is a minimum SCTF of S_p. Hence, $\tau_C^-(S_p) = 0$ and $\tau_C^s(S_p) = 0$. The proof for $\tau_C^-(G) = 0$ is analogous to that for $\tau_C^s(G) = 0$. □

Theorem 1 (Lee and Chang [9]). *Let S_p be a sun. Then,*

(1) $\gamma^-(S_p) = \gamma_s(S_p) = 0$;
(2) $\gamma_{\{k\}}(S_p) = \lceil pk/2 \rceil$;
(3) $\gamma_{\times 1}(S_p) = \lceil p/2 \rceil, \gamma_{\times 2}(S_p) = p$ and $\gamma_{\times 3}(S_p) = 2p$.

Corollary 1. *Let S_p be a sun. Then,*

(1) $\gamma^-(S_p) = \tau_C^-(S_p) = \gamma_s(S_p) = \tau_C^s(S_p) = 0$;
(2) $\gamma_{\{k\}}(S_p) = \tau_C^{\{k\}}(S_p) = \lceil pk/2 \rceil$ for $k > 1$;
(3) $\gamma_{\times 2}(S_p) = \tau_C^2(S_p) = p$ and $\gamma_{\times 3}(S_p) = \tau_C^3(S_p) = 2p$.

Proof. The corollary holds by Lemmas 1–3 and Corollary 1. □

3. Clique Perfect Graphs

Let $\mathcal{G}$ be the set of all induced subgraphs of G. If $\tau_C(H) = \alpha_C(H)$ for every $H \in \mathcal{G}$, then G is *clique perfect*. In this section, we study the $\{k\}$-CTP for clique perfect graphs and the SCTP for balanced graphs.

Lemma 4. *Let G be such a graph that $\tau_C(G) = \alpha_C(G)$. Then, $\tau_C^{\{k\}}(G) = k\tau_C(G)$.*

Proof. Assume that D is a minimum CTS of G. Then, $|D| = \tau_C(G)$. Let $x \in V(G)$ and let f be a function whose domain is $V(G)$ and range is $\{0, 1, \ldots, k\}$, and $f(x) = k$ if $x \in D$; otherwise, $f(x) = 0$. Clearly, f is a $\{k\}$-CTF of G. We have $\tau_C^{\{k\}}(G) \le k\tau_C(G)$.

Assume that f is a minimum-weight $\{k\}$-CTF of G. Then, $f(V(G)) = \tau_C^{\{k\}}(G)$ and $f(C) \ge k$ for every $C \in C(G)$. Let $S = \{C_1, C_2, \ldots, C_\ell\}$ be a maximum CIS of G. We know that $|S| = \ell = \alpha_C(G)$ and $\sum_{i=1}^{\ell} f(C_i) \le f(V(G))$. Therefore, $k\tau_C(G) = k\alpha_C(G) = k\ell \le \sum_{i=1}^{\ell} f(C_i) \le f(V(G)) = \tau_C^{\{k\}}(G)$. Following what we have discussed above, we know that $\tau_C^{\{k\}}(G) = k\tau_C(G)$. □

Theorem 2. *If a graph G is clique perfect, $\tau_C^{\{k\}}(G) = k\tau_C(G)$.*

Proof. Since G is clique perfect, $\tau_C(G) = \alpha_C(G)$. Hence, the theorem holds by Lemma 4. □

Corollary 2. *The $\{k\}$-CTP is polynomial-time solvable for distance-hereditary graphs, balanced graphs, strongly chordal graphs, comparability graphs, and chordal graphs without odd suns.*

Proof. Distance-hereditary graphs, balanced graphs, strongly chordal graphs, comparability graphs, and chordal graphs without odd suns are clique perfect, and the CTP can be solved in polynomial time for them [10–14]. The corollary therefore holds. □

Definition 8. *Suppose that R is a function whose domain is $C(G)$ and range is $\{0, 1, \ldots, \omega(G)\}$. If $R(C) \le |C|$ for every $C \in C(G)$, then R is a clique-size restricted function (abbreviated as*

CSRF) of G. A set $D \subseteq V(G)$ is an *R-clique transversal set* (abbreviated as *R-CTS*) of G if R is a CSRF of G and $|D \cap C| \geq R(C)$ for every $C \in C(G)$. Let $\tau_R(G) = min\{|D| \mid D$ is an R-CTS of $G\}$. The generalized clique transversal problem (abbreviated as GCTP) is to find a minimum R-CTS for a graph G with a CSRF R.

Lemma 5. *Let G be a graph with a CSRF R. If $R(C) = \lceil (|C| + 1)/2 \rceil$ for every $C \in C(G)$, then $\tau_C^s(G) = 2\tau_R(G) - n$.*

Proof. Assume that D is a minimum R-CTS of G. Then, $|D| = \tau_R(G)$. Let $x \in V(G)$ and let f be a function of G whose domain is $V(G)$ and range is $\{-1, 1\}$, and $f(x) = 1$ if $x \in D$; otherwise, $f(x) = -1$. Since $|D \cap C| \geq \lceil (|C| + 1)/2 \rceil$ for every $C \in C(G)$, there are at least $\lceil (|C| + 1)/2 \rceil$ vertices in C with the function value 1. Therefore, $f(C) \geq 1$ for every $C \in C(G)$, and f is an SCTF of G. Then, $\tau_C^s(G) \leq 2\tau_R(G) - n$.

Assume that h is a minimum-weight SCTF of G. Clearly, $h(V(G)) = \tau_C^s(G)$. Since $h(C) \geq 1$ for every $C \in C(G)$, C contains at least $\lceil (|C| + 1) \rceil / 2$ vertices with the function value 1. Let $D = \{x \mid h(x) = 1, x \in V(G)\}$. The set D is an R-CTS of G. Therefore, $2\tau_R(G) - n \leq 2|D| - n = \tau_C^s(G)$. Hence, we have $\tau_C^s(G) = 2\tau_R(G) - n$. □

Theorem 3. *The SCTP on balanced graphs can be solved in polynomial time.*

Proof. Suppose that a graph G has n vertices $v_1, v_2, \ldots, v_n$ and ℓ maximal cliques $C_1, C_2, \ldots, C_\ell$. Let $i \in \{1, 2, \ldots, \ell\}$ and $j \in \{1, 2, \ldots, n\}$. Let M be an $\ell \times n$ matrix such that an element $M(i, j)$ of M is one if the maximal clique C_i contains the vertex v_j, and $M(i, j) = 0$ otherwise. We call M the *clique matrix* of G. If the clique matrix M of G does not contain a square submatrix of odd order with exactly two ones per row and column, then M is a *balanced* matrix and G is a *balanced* graph. We formulae the GCTP on a balanced graph G with a CSRF R as the following integer programming problem:

$$\left. \begin{array}{ll} \text{minimize} & \sum_{i=1}^{n} x_i \\[2mm] \text{subject to} & MX \geq C \end{array} \right\}$$

where $C = (R(C_1), R(C_2), \ldots, R(C_\ell))$ is a column vector and $X = (x_1, x_2, \ldots, x_n)$ is a column vector such that x_i is either 0 or 1. Since the matrix M is balanced, an optimal 0–1 solution of the integer programming problem above can be found in polynomial time by the results in [15]. By Lemma 5, we know that the SCTP on balanced graphs can be solved in polynomial time. □

4. Split Graphs

Let G be such a graph that $V(G) = I \cup C$ and $I \cap C = \varnothing$. If I is an independent set and C is a clique, G is a *split* graph. Then, every maximal of G is either C itself, or the closed neighborhood $N_G[x]$ of a vertex $x \in I$. We use $G = (I, C, E)$ to represent a split graph. The $\{k\}$-CTP, the k-FCTP, the SCTP, and the MCTP for split graphs are considered in this section. We also consider the $\{k\}$-DP, the k-TDP, the SDP, and the MDP for split graphs.

For split graphs, the $\{k\}$-DP, the k-TDP, and the MDP are NP-complete [16–18], but the complexity of the SDP is still unknown. In the following, we examine the relationships between the $\{k\}$-CTP and the $\{k\}$-DP, the k-FCTP and the k-TDP, the SCTP and the SDP, and the MCTP and the MDP. Then, by the relationships, we prove the NP-completeness of the SDP, the $\{k\}$-CTP, the k-FCTP, the SCTP, and the MCTP for split graphs. We first consider the $\{k\}$-CTP and the k-FCTP and show in Theorems 4 and 5 that $\tau_C^k(G) = \gamma_{\times k}(G)$ and $\tau_C^{\{k\}}(G) = \gamma_{\{k\}}(G)$ for any split graph G. Chordal graphs form a superclass of split graphs [19]. The cardinality of $C(G)$ is at most n for any chordal graph G [20]. The following lemma therefore holds trivially.

Lemma 6. *The k-FCTP, the $\{k\}$-CTP, the SCTP, and the MCTP for chordal graphs are in NP.*

Theorem 4. *Suppose that $k \in \mathbb{N}$ and $G = (I, C, E)$ is a split graph. Then, $\tau_C^k(G) = \gamma_{\times k}(G)$.*

Proof. Let S be a minimum k-FCTS of G. Consider a vertex $y \in I$. By the structure of G, $N_G[y]$ is a maximal clique of G. Then, $|S \cap N_G[y]| \geq k$. We now consider a vertex $x \in C$. If $C \notin C(G)$, then there exists a vertex $y \in I$ such that $N_G[y] = C \cup \{y\}$. Clearly, $N_G[y] \subseteq N_G[x]$ and $|S \cap N_G[x]| \geq |S \cap N_G[y]| \geq k$. If $C \in C(G)$, then $|S \cap N_G[x]| \geq |S \cap C| \geq k$. Hence, S is a k-TDS of G. We have $\gamma_{\times k}(G) \leq \tau_C^k(G)$.

Let D be a minimum k-TDS of G. Recall that the closed neighborhood of every vertex in I is a maximal clique. Then, D contains at least k vertices in the maximal clique $N_G[y]$ for every vertex $y \in I$. If $C \notin C(G)$, D is clearly a k-FCTS of G. Suppose that $C \in C(G)$. We consider three cases as follows.

Case 1: $y \in I \setminus D$. Then, $|D \cap C| \geq |D \cap N_G(y)| \geq k$. The set D is a k-FCTS of G.

Case 2: $y \in I \cap D$ and $x \in N_G(y) \setminus D$. Then, the set $D' = (D \setminus \{y\}) \cup \{x\}$ is still a minimum k-TDS and $|D' \cap C| \geq |D' \cap N_G(y)| \geq k$. The set D' is a k-FCTS of G.

Case 3: $I \subseteq D$ and $N_G[y] \subseteq D$ for every $y \in I$. Then, $|D \cap C| \geq |D \cap N_G(y)| \geq k - 1$. Since $C \in C(G)$, there exists $x \in C$ such that $y \notin N_G(x)$. If $N_G(x) \cap I = \emptyset$, then $N_G[x] = C$ and $|D \cap C| = |D \cap N_G[x]| \geq k$. Otherwise, let $y' \in N_G(x) \cap I$. Then, $x \in D$ and $|D \cap C| \geq |D \cap N_G(y)| + 1 \geq k$. The set D is a k-FCTS of G.

By the discussion of the three cases, we have $\tau_C^k(G) \leq \gamma_{\times k}(G)$. Hence, we obtain that $\gamma_{\times k}(G) \leq \tau_C^k(G)$ and $\tau_C^k(G) \leq \gamma_{\times k}(G)$. The theorem holds for split graphs. □

Theorem 5. *Suppose that $k \in \mathbb{N}$ and $G = (I, C, E)$ is a split graph. Then, $\tau_C^{\{k\}}(G) = \gamma_{\{k\}}(G)$.*

Proof. We can verify by contradiction that G has a minimum-weight $\{k\}$-CTF f and a minimum-weight $\{k\}$-DF g of G such that $f(y) = 0$ and $g(y) = 0$ for every $y \in I$. By the structure of G, $N_G[y] \in C(G)$ for every $y \in I$. Then, $f(N_G[y]) \geq k$ and $g(N_G[y]) \geq k$. Since $f(y) = 0$ and $g(y) = 0$, $f(N_G(y)) \geq k$ and $g(N_G(y)) \geq k$.

For every $y \in I$, $N_G(y) \subseteq C$ and $f(C) \geq f(N_G(y)) \geq k$. For every $x \in C$, $f(N_G[x]) \geq f(C) \geq k$. Therefore, the function f is also a $\{k\}$-DF of G. We have $\gamma_{\{k\}}(G) \leq \tau_C^{\{k\}}(G)$. We now consider $g(C)$ for the clique C. If $C \notin C(G)$, the function g is clearly a $\{k\}$-CTF of G. Suppose that $C \in C(G)$. Notice that g is a $\{k\}$-DF and $g(y) = 0$ for every $y \in I$. Then, $g(C) = g(N_G[x]) \geq k$ for any vertex $x \in C$. Therefore, g is also a $\{k\}$-CTF of G. We have $\tau_C^{\{k\}}(G) \leq \gamma_{\{k\}}(G)$. Following what we have discussed above, we know that $\tau_C^{\{k\}}(G) = \gamma_{\{k\}}(G)$. □

Corollary 3. *The $\{k\}$-CTP and the k-FCTP are NP-complete for split graphs.*

Proof. The corollary holds by Theorems 4 and 5 and the NP-completeness of the $\{k\}$-DP and the k-TDP for split graphs [16,18]. □

A graph G is a *complete* if $C(G) = \{V(G)\}$. Let G be a complete graph and let $x \in V(G)$. The vertex set $V(G)$ is the union of the sets $\{x\}$ and $V(G) \setminus \{x\}$. Clearly, $\{x\}$ is an independent set and $V(G) \setminus \{x\}$ is a clique of G. Therefore, complete graphs are split graphs. It is easy to verify the Lemma 7.

Lemma 7. *If G is a complete graph and $k \in \mathbb{N}$, then*

(1) $\tau_C^k(G) = \gamma_{\times k}(G) = k$ *for $k \leq n$;*

(2) $\tau_C^{\{k\}}(G) = \gamma_{\{k\}}(G) = k$;

(3) $\tau_C^-(G) = \gamma^-(G) = 1$;

(4) $\tau_C^s(G) = \gamma_s(G) = \begin{cases} 1 & \text{if } n \text{ is odd;} \\ 2 & \text{otherwise.} \end{cases}$

For split graphs, however, the signed and minus domination numbers are not necessarily equal to the signed and minus clique transversal numbers, respectively. Figure 2 shows a split graph G with $\tau_C^s(G) = \tau_C^-(G) = -3$. However, $\gamma_s(G) = \gamma^-(G) = 1$. We therefore introduce H_1-*split* graphs and show in Theorem 6 that their signed and minus domination numbers are equal to the signed and minus clique transversal numbers, respectively. H_1-split graphs are motivated by the graphs in [17] for proving the NP-completeness of the MDP on split graphs. Figure 3 shows an H_1-split graph.

$$-1 \quad -1 \quad -1 \quad -1 \quad -1$$

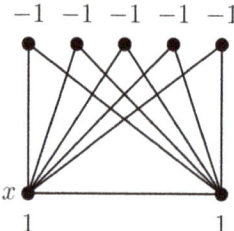

$$x$$
$$1 \qquad\qquad 1$$

Figure 2. A split graph G with $\tau_C^s(G) = \tau_C^-(G) = -3$.

Definition 9. *Suppose that $G = (I, C, E)$ is a split graph with $3p + 3\ell + 2$ vertices. Let U, S, X, and Y be pairwise disjoint subsets of $V(G)$ such that $U = \{u_i \mid 1 \le i \le p\}$, $S = \{s_i \mid 1 \le i \le \ell\}$, $X = \{x_i \mid 1 \le i \le p + \ell + 1\}$, and $Y = \{y_i \mid 1 \le i \le p + \ell + 1\}$. The graph G is an H_1-split graph if $V(G) = U \cup S \cup X \cup Y$ and G entirely satisfies the following three conditions.*

(1) $I = S \cup Y$ and $C = U \cup X$.
(2) $N_G(y_i) = \{x_i\}$ for $1 \le i \le p + \ell + 1$.
(3) $|N_G(s_i) \cap U| = 3$ and $N_G(s_i) \cap X = \{x_i\}$ for $1 \le i \le \ell$.

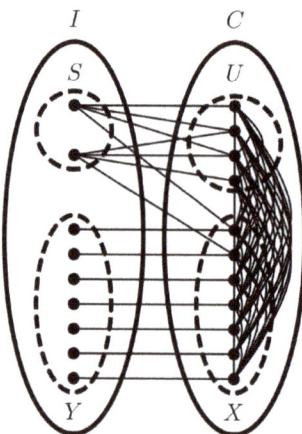

Figure 3. A split graph G with one of its partitions indicated.

Theorem 6. *For any H_1-split graph $G = (I, C, E)$, $\tau_C^s(G) = \gamma_s(G)$ and $\tau_C^-(G) = \gamma^-(G)$.*

Proof. We first prove $\tau_C^s(G) = \gamma_s(G)$. Let $G = (I, C, E)$ be an H_1-split graph. As stated in Definition 9, I can be partitioned into $S = \{s_i \mid 1 \le i \le \ell\}$ and $Y = \{y_i \mid 1 \le i \le p + \ell + 1\}$, and C can be partitioned into $U = \{u_i \mid 1 \le i \le p\}$ and $X = \{x_i \mid 1 \le i \le p + \ell + 1\}$. Assume that f is a minimum-weight SDF of G. For each $y_i \in Y$, $|N_G[y_i]| = 2$ and y_i is adjacent to only the vertex $x_i \in X$. Then, $f(x_i) = f(y_i) = 1$ for $1 \le i \le p + \ell + 1$. Since $C = U \cup X$ and $|U| = p$, we know that $f(C) = f(U) + f(X) \ge (-p) + (p + \ell + 1) \ge \ell + 1$.

Notice that $f(N_G[y]) \geq 1$ and $N_G[y] \in C(G)$ for every $y \in I$. Therefore, f is also an SCTF of G. We have $\tau_C^s(G) \leq \gamma_s(G)$.

Assume that h is a minimum-weight SCTF of G. For each $y_i \in Y$, $|N_G[y_i]| = 2$ and y_i is adjacent to only the vertex $x_i \in X$. Then, $h(x_i) = h(y_i) = 1$ for $1 \leq i \leq p + \ell + 1$. Consider the vertices in I. Since $N_G[y] \in C(G)$ for every $y \in I$, $h(N_G[y]) \geq 1$. We now consider the vertices in C. Recall that $C = U \cup X$. Let $u_i \in U$. Since $|U| = p$ and $|S| = \ell$, we know that $h(N_G[u_i]) = h(U) + h(X) + h(N_G[u_i] \cap S) \geq (-p) + (p + \ell + 1) + (-\ell) \geq 1$. Let $x_i \in X$. Then, $h(N_G[x_i]) = h(U) + h(X) + h(y_i) + h(s_i) \geq (-p) + (p + \ell + 1) + 1 - 1 \geq \ell + 1$. Therefore, h is also an SDF of G. We have $\gamma_s(G) \leq \tau_C^s(G)$.

Following what we have discussed above, we have $\tau_C^s(G) = \gamma_s(G)$. The proof for $\tau_C^-(G) = \gamma^-(G)$ is analogous to that for $\tau_C^s(G) = \gamma_s(G)$. Hence, the theorem holds for any H_1-split graphs. □

Theorem 7. *The SDP on H_1-split graphs is NP-complete.*

Proof. We reduce the (3,2)-*hitting set problem* to the SDP on H_1-split graphs. Let $U = \{u_i \mid 1 \leq i \leq p\}$ and let $\mathcal{C} = \{C_1, C_2, \ldots, C_\ell\}$ such that $C_i \subseteq U$ and $|C_i| = 3$ for $1 \leq i \leq \ell$. A (3,2)-hitting set for the instance $(U, \mathcal{C})$ is a subset U' of U such that $|C_i \cap U'| \geq 2$ for $1 \leq i \leq \ell$. The (3,2)-hitting set problem is to find a minimum (3,2)-hitting set for any instance $(U, \mathcal{C})$. The (3,2)-hitting set problem is NP-complete [17].

Consider an instance $(U, \mathcal{C})$ of the (3,2)-hitting set problem. Let $S = \{s_i \mid 1 \leq i \leq \ell\}$, $X = \{x_i \mid 1 \leq i \leq p + \ell + 1\}$, and $Y = \{y_i \mid 1 \leq i \leq p + \ell + 1\}$. We construct an H_1-split graph $G = (I, C, E)$ by the following steps.

(1)   Let $I = S \cup Y$ be an independent set and let $C = U \cup X$ be a clique.
(2)   For each vertex $s_i \in S$, a vertex $u \in U$ is connected to s_i if $u \in C_i$.
(3)   For $1 \leq i \leq p + \ell + 1$, the vertex y_i is connected to the vertex x_i.
(4)   For $1 \leq i \leq \ell$, the vertex s_i is connected to the vertex x_i.

Let $\tau_h(3, 2)$ be the minimum cardinality of a (3,2)-hitting set for the instance $(U, \mathcal{C})$. Assume that U' is a minimum (3,2)-hitting set for the instance $(U, \mathcal{C})$. Then, $|U'| = \tau_h(3, 2)$. Let f be a function whose domain is $V(G)$ and range is $\{-1, 1\}$, and $f(v) = 1$ if $v \in X \cup Y \cup U'$ and $f(v) = -1$ if $v \in S \cup (U \setminus U')$. Clearly, f is an SDF of G. We have $\gamma_s(G) \leq 2(p + \ell + 1) + |U'| - \ell - (p - |U'|) = p + \ell + 2\tau_h(3, 2) + 2$.

Assume that f is minimum-weight SDF of G. For each $y_i \in Y$, $|N_G[y_i]| = 2$ and y_i is adjacent to only the vertex $x_i \in X$. Then, $f(x_i) = f(y_i) = 1$ for $1 \leq i \leq p + \ell + 1$. For any vertex $v \in X \cup Y \cup U$, $f(N_G[v]) \geq 1$ no matter what values the function f assigns to the vertices in U or in S. By the construction of G, $deg_G(s_i) = 4$ and $|N_G[s_i]| = 5$ for $1 \leq i \leq \ell$. There are at least three vertices in $N_G[s_i]$ with the function value 1. If $f(N_G[s_i]) = 5$, then there exists an SDF g of G such that $g(s_i) = -1$ and $g(v) = f(v)$ for every $v \in V(G) \setminus \{s_i\}$. Then, $g(V(G)) < f(V(G))$. It contradicts the assumption that the weight of f is minimum. Therefore, there exists a minimum-weight SDF h of G such that $h(s_i) = -1$ for $1 \leq i \leq \ell$. Notice that $N_G(s_i) = C_i \cup \{x_i\}$ for $1 \leq i \leq \ell$. There are at least two vertices in C_i with the function value 1. Then, the set $U' = \{u \in U \mid h(u) = 1\}$ is a (3,2)-hitting set for the instance $(U, \mathcal{C})$. We have $p + \ell + 2\tau_h(3, 2) + 2 \leq p + \ell + 2|U'| + 2 = \gamma_s(G)$.

Following what we have discussed above, we know that $\gamma_s(G) = p + \ell + 2\tau_h(3, 2) + 2$. Hence, the SDP on H_1-split graphs is NP-complete. □

Corollary 4. *The SCTP and the MCTP on split graphs are NP-complete.*

Proof. The corollary holds by Theorems 6 and 7 and the NP-completeness of the MDP on split graphs [17]. □

5. Doubly Chordal and Dually Chordal Graphs

Assume that G is a graph with n vertices $x_1, x_2, \ldots, x_n$. Let $i \in \{1, 2, \ldots, n\}$ and let G_i be the subgraph $G[V(G) \setminus \{x_1, x_2, \ldots x_{i-1}\}]$. For every $x \in V(G_i)$, let $N_i[x] = \{y \mid y \in (N_G[x] \setminus \{x_1, x_2, \ldots, x_{i-1}\})\}$. In G_i, if there exists a vertex $x_j \in N_i[x_i]$ such that $N_i[x_k] \subseteq N_i[x_j]$ for every $x_k \in N_i[x_i]$, then the ordering $(x_1, x_2, \ldots, x_n)$ is a *maximum neighborhood ordering* (abbreviated as MNO) of G. A graph G is *dually chordal* [21] if and only if G has an MNO. It takes linear time to compute an MNO for any dually chordal graph [22]. A graph G is a *doubly chordal* graph if G is both chordal and dually chordal [23]. Lemma 8 shows that a dually chordal graph is not necessarily a chordal graph or a clique perfect graph. Notice that the number of maximal cliques in a chordal graph is at most n [20], but the number of maximal cliques in a dually chordal graph can be exponential [24].

Lemma 8. *For any dually graph G, $\tau_C(G) = \alpha_C(G)$, but G is not necessarily clique perfect or chordal.*

Proof. Brandstädt et al. [25] showed that the CTP is a particular case of the *clique r-domination problem* and the CIP is a particular case of the *clique r-packing problem*. They also showed that the minimum cardinality of a clique r-dominating set of a dually chordal graph G is equal to the maximum cardinality of a clique r-packing set of G. Therefore, $\tau_C(G) = \alpha_C(G)$.

Assume that H is a graph obtained by connecting every vertex of a cycle C_4 of four vertices x_1, x_2, x_3, x_4 to a vertex x_5. Clearly, the ordering $(x_1, x_2, x_3, x_4, x_5)$ is an MNO and thus H is a dually chordal graph. The cycle C_4 is an induced subgraph of H and does not have a chord. Moreover, $\tau_C(H) = \alpha_C(H) = 1$, but $\tau_C(C_4) = 2$ and $\alpha_C(C_4) = 1$. Hence, a dually chordal graph is not necessarily clique perfect or chordal. □

Theorem 8. *Suppose that $k \in \mathbb{N}$ and $k > 1$. The k-FCTP on doubly chordal graphs is NP-complete.*

Proof. Suppose that G is a chordal graph. Let H be a graph such that $V(H) = V(G) \cup \{x\}$ and $E(H) = E(G) \cup \{(x, y) \mid y \in V(G)\}$. Clearly, H is a doubly chordal graph and we can construct H from G in linear time.

Assume that S is a minimum $(k-1)$-FCTS of G. By the construction of H, each maximal clique of H contains the vertex x. Therefore, $S \cup \{x\}$ is a k-FCTS of H. Then $\tau_C^k(H) \leq \tau_C^{k-1}(G) + 1$.

By contradiction, we can verify that there exists a minimum k-FCTS D of H such that $x \in D$. Let $S = D \setminus \{x\}$. Clearly, S is a $(k-1)$-FCTS of G. Then $\tau_C^{k-1}(G) \leq \tau_C^k(H) - 1$. Following what we have discussed above, we have $\tau_C^k(H) = \tau_C^{k-1}(G) + 1$. Notice that $\tau_C(G) = \tau_C^1(G)$ and the CTP on chordal graphs is NP-complete [14]. Hence, the k-FCTP on doubly chordal graphs is NP-complete for doubly chordal graphs. □

Theorem 9. *For any doubly chordal graph G, $\tau_C^{\{k\}}(G)$ can be computed in linear time.*

Proof. The clique r-dominating problem on doubly chordal graphs can be solved in linear time [25]. The CTP is a particular case of the clique r-domination problem. Therefore, the CTP on doubly chordal graphs can also be solved in linear time. By Lemmas 4 and 8, the theorem holds. □

6. k-Trees

Assume that G is a graph with n vertices $x_1, x_2, \ldots, x_n$. Let $i \in \{1, 2, \ldots, n\}$ and let G_i be the subgraph $G[V(G) \setminus \{x_1, x_2, \ldots x_{i-1}\}]$. For every $x \in V(G_i)$, let $N_i[x] = \{y \mid y \in (N_G[x] \setminus \{x_1, x_2, \ldots, x_{i-1}\})\}$. If $N_i[x_i]$ is a clique for $1 \leq i \leq n$, then the ordering $(x_1, x_2, \ldots, x_n)$ is a *perfect elimination ordering* (abbreviated as PEO) of G. A graph G is chordal if and only if G has a PEO [26]. A chordal graph G is a k-tree if and only if either G is a complete graph of k vertices or G has more than k vertices and there exists a PEO

$(x_1, x_2, \ldots, x_n)$ such that $N_i[x_i]$ is a clique of k vertices if $i = n - k + 1$; otherwise, $N_i[x_i]$ is a clique of $k + 1$ vertices for $1 \leq i \leq n - k$. Figure 4 shows a 2-tree with the PEO $(v_1, v_2, \ldots, v_{13})$.

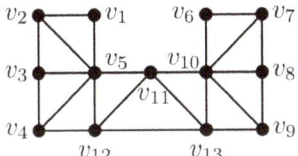

Figure 4. A 2-tree H.

In [3], Chang et al. showed that the MCTP is NP-complete for k-trees with unbounded k by proving $\gamma(G) = \tau_M(G)$ for any k-tree G. However, Figure 4 shows a counterexample that disproves $\gamma(G) = \tau_M(G)$ for any k-tree G. The graph H in Figure 4 is a 2-tree with the perfect elimination ordering $(v_1, v_2, \ldots, v_{13})$. The set $\{v_5, v_{10}\}$ is the minimum dominating set of H and the set $\{v_5, v_{10}, v_{11}\}$ is a minimum MCTS of H. A modified NP-completeness proof is therefore desired for the MCTP on k-tree with unbounded k.

Theorem 10. *The MCTP and the MCIP are NP-complete for k-trees with unbounded k.*

Proof. The CTP and the CIP are NP-complete for k-trees with unbounded k [8]. Since every maximal clique in a k-tree is also a maximum clique [27], an MCTS is a CTS and an MCIS is a CIS. Hence, the MCTP and the MCIP are NP-complete for k-trees with unbounded k. □

Theorem 11. *The SCTP is NP-complete for k-trees with unbounded k.*

Proof. Suppose that $k_1 \in \mathbb{N}$ and G is a k_1-tree with $|V(G)| > k_1$. Let $C(G) = \{C_1, C_2, \ldots, C_\ell\}$. Since G is a k_1-tree, $|C_i| = k_1 + 1$ for $1 \leq i \leq \ell$.

Let Q be a clique with $k_1 + 1$ vertices. Let H be a graph such that $V(H) = V(G) \cup Q$ and $E(H) = E(G) \cup \{(x, y) \mid x, y \in Q\} \cup \{(x, y) \mid x \in Q, y \in V(G)\}$. Let $X_i = C_i \cup Q$ be a clique for $1 \leq i \leq \ell$. Clearly, $C(H) = \{X_i \mid 1 \leq i \leq \ell\}$. Let $k_2 = 2k_1 + 1$. Then, H is a k_2-tree and $|X_i| = k_2 + 1 = 2k_1 + 2$ for $1 \leq i \leq \ell$. Clearly, we can verify that there exists a minimum-weight SCTF h of H of such that $h(x) = 1$ for every $x \in Q$. Then, $C_i = X_i \setminus Q$ contains at least one vertex x with $h(x) = 1$ for $1 \leq i \leq \ell$. Let $S = \{x \mid x \in V(H) \setminus Q$ and $h(x) = 1\}$. Then, S is a CTS of G. Since $\tau_C^s(H) = |Q| + 2|S| - |V(G)|$, we have $|Q| + 2\tau_C(G) - |V(G)| \leq \tau_C^s(H)$.

Assume that D is a minimum CTS of G. Let f be a function of H whose domain is $V(H)$ and range is $\{-1, 1\}$, and (1) $f(x) = 1$ for every $x \in Q$, (2) $f(x) = 1$ for every $x \in D$, and (3) $f(x) = -1$ for every $x \in V(G) \setminus D$. Each maximal clique of H has at least $k_1 + 2$ vertices with the function value 1. Therefore, f is an SCTF. We have $\tau_C^s(H) \leq |Q| + 2\tau_C(G) - |V(G)|$. Following what we have discussed above, we know that $\tau_C^s(H) = |Q| + 2\tau_C(G) - |V(G)|$. The theorem therefore holds by the NP-completeness of the CTP for k-trees [8]. □

Theorem 12. *Suppose that $\kappa \in \mathbb{N}$ the κ-FCTP is NP-complete on k-trees with unbounded k.*

Proof. Assume that $k_1 \in \mathbb{N}$ and G is a k_1-tree with $|V(G)| > k_1$. Let H be a graph such that $V(H) = V(G) \cup \{x\}$ and $E(H) = E(G) \cup \{(x, y) \mid y \in V(G)\}$. Clearly, H is a $(k_1 + 1)$-tree and we can construct H in linear time. Following the argument analogous to the proof of Theorem 8, we have $\tau_C^\kappa(H) = \tau_C^{\kappa-1}(G) + 1$. The theorem therefore holds by the NP-completeness of the CTP for k-trees [8]. □

Theorem 13. *The SCTP and κ-FCTP problems can be solved in linear-time for k-trees with fixed k.*

Proof. Assume that $\kappa \in \mathbb{N}$ and G is a graph. The κ-FCTP is the GCTP with the CSRF R whose domain is $C(G)$ and range is $\{\kappa\}$. By Lemma 5, $\tau_C^s(G)$ can be obtained from the solution to the GCTP on a graph G with a particular CSRF R. Since the GCTP is linear-time solvable for k-trees with fixed k [8], the SCTP and κ-FCTP are also linear-time solvable for k-trees with fixed k. $\square$

7. Planar, Total, and Line Graphs

In a graph, a vertex x and an edge e are *incident* to each other if e connects x to another vertex. Two edges are *adjacent* if they share a vertex in common. Let G and H be graphs such that each vertex $x \in V(H)$ corresponds to an edge $e_x \in E(G)$ and two vertices $x, y \in V(H)$ are adjacent in H if and only if their corresponding edges e_x and e_y are adjacent in G. Then, H is the *line graph* of G and denoted by $L(G)$. Let H' be a graph such that $V(H') = V(G) \cup E(G)$ and two vertices $x, y \in V(H')$ are adjacent in H if and only if x and y are adjacent or incident to each other in G. Then, H' is the *total graph* of G and denoted by $T(G)$.

Lemma 9 ([28]). *The following statements hold for any triangle-free graph G.*

(1) *Every maximal clique of $L(G)$ is the set of edges of G incident to some vertex of G.*
(2) *Two maximal cliques in $L(G)$ intersect if and only if their corresponding vertices (in G) are adjacent in G.*

Theorem 14. *The MCIP is NP-complete for any 4-regular planar graph G with the clique number 3.*

Proof. Since $|C(G)| = O(n)$ for any planar graph G [29], the MCIP on planar graphs is in NP. Let $\mathcal{G}$ be the class of triangle-free, 3-connected, cubic planar graphs. The independent set problem remains NP-complete even when restricted to the graph class $\mathcal{G}$ [30]. We reduce this NP-complete problem to the MCIP for 4-regular planar graphs with the clique number 3 as follows.

Let $G \in \mathcal{G}$ and $H = L(G)$. Clearly, we can construct H in polynomial time. By Lemma 9, we know that H is a 4-regular planar graph with $\omega(H) = 3$ and each maximal clique is a triangle in H.

Assume that $D = \{x_1, x_2, \ldots, x_\ell\}$ is an independent set of G of maximum cardinality. Since $G \in \mathcal{G}$, $deg_G(x) = 3$ for every $x \in V(G)$. Let $e_{i_1}, e_{i_2}, e_{i_3} \in E(G)$ have the vertex x_i in common for $1 \le i \le \ell$. Then, $e_{i_1}, e_{i_2}, e_{i_3}$ form a triangle in H. Let C_i be the triangle formed by $e_{i_1}, e_{i_2}, e_{i_3}$ in H for $1 \le i \le \ell$. For each pair of vertices $x_j, x_k \in D$, x_j is not adjacent to x_k in G. Therefore, C_j and C_k in H do not intersect. The set $\{C_1, C_2, \ldots, C_\ell\}$ is an MCIS of H. We have $\alpha(G) \le \alpha_M(H)$.

Assume that $S = \{C_1, C_2, \ldots, C_\ell\}$ is a maximum MCIS of H. Then, each $C_i \in S$ is a triangle in H. Let C_i be formed by $e_{i_1}, e_{i_2}, e_{i_3}$ in H for $1 \le i \le \ell$. Then, $e_{i_1}, e_{i_2}, e_{i_3}$ are incident to the same vertex in G. For $1 \le i \le \ell$, let $e_{i_1}, e_{i_2}, e_{i_3} \in E(G)$ have the vertex x_i in common. For each pair of $C_j, C_k \in S$, C_j and C_k do not intersect. Therefore, x_j is not adjacent to x_k in G. The set $\{x_1, x_2, \ldots, x_\ell\}$ is an independent set of G. We have $\alpha_M(H) \le \alpha(G)$.

Hence, $\alpha(G) = \alpha_M(H)$. For $k \in \mathbb{N}$, we know that $\alpha(G) \ge k$ if and only if $\alpha_M(G) \ge k$. $\square$

Corollary 5. *The MCIP is NP-complete for line graphs of triangle-free, 3-connected, cubic planar graphs.*

Proof. The corollary holds by the reduction of Theorem 14. $\square$

Theorem 15. *The MCIP problem is NP-complete for total graphs of triangle-free, 3-connected, cubic planar graphs.*

Proof. Since $|C(G)| = O(n)$ for a planar graph G, the MCIP on planar graphs is in NP. Let $\mathcal{G}$ be the classes of traingle-free, 3-connected, cubic planar graphs. The independent set problem remains NP-complete even when restricted to the graph class $\mathcal{G}$ [30]. We reduce this NP-complete problem to MCIP for for total graphs of triangle-free, 3-connected, cubic planar graphs. as follows

Let $G \in \mathcal{G}$ and $H = T(G)$. Clearly, we can construct H in polynomial time. By Lemma 9, we can verify that H is a 6-regular graph with $\omega(H) = 4$.

Assume that $D = \{x_1, x_2, \ldots, x_\ell\}$ is an independent set of G of maximum cardinality. Since $G \in \mathcal{G}$, $deg_G(x) = 3$ for every $x \in V(G)$. Let $e_{i_1}, e_{i_2}, e_{i_3} \in E(G)$ have the vertex x_i in common. Then, $x_i, e_{i_1}, e_{i_2}, e_{i_3}$ form a maximum clique in H. Let C_i be the maximum clique formed by $x_i, e_{i_1}, e_{i_2}, e_{i_3}$ in H for $1 \le i \le \ell$. For each pair of vertices $x_j, x_k \in D$, x_j is not adjacent to x_k in G. Therefore, C_j and C_k in H do not intersect. The set $\{C_1, C_2, \ldots, C_\ell\}$ is an MCIS of H. We have $\alpha(G) \le \alpha_M(H)$.

Assume that $S = \{C_1, C_2, \ldots, C_\ell\}$ is a maximum MCIS of H. By the construction of H, each $C_i \in S$ is formed by three edge-vertices in $E(G)$ and their common end vertex in $V(G)$. Let $x_i \in V$ and $e_{i_1}, e_{i_2}, e_{i_3} \in E(G)$ in H such that C_i is formed by $v_i, e_{i_1}, e_{i_2}, e_{i_3}$ for $1 \le i \le \ell$. For each pair of $C_j, C_k \in C$, C_j and C_k do not intersect. Therefore, x_j is not adjacent to x_k in G. The set $\{x_1, x_2, \ldots, x_\ell\}$ is an independent set of G. We have $\alpha_M(H) \le \alpha(G)$.

Hence, $\alpha(G) = \alpha_M(H)$. For $k \in \mathbb{N}$, we know that $\alpha(G) \ge k$ if and only if $\alpha_M(H) \ge k$. $\square$

Funding: This research is supported by a Taiwanese grant under Grant No. NSC-97-2218-E-130-002-MY2.

Acknowledgments: We are grateful to the anonymous referees for their valuable comments and suggestions to improve the presentation of this paper.

Conflicts of Interest: The author declares no conflict of interest.

References

1. Dahlhaus, E.; Kratochvíl, J.; Manuel, P.D.; Miller, M. Transversal partitioning in balanced hypergraphs. *Discret. Math.* **1997**, *79*, 75–89. [CrossRef]
2. Dahlhaus, E.; Manuel, P.D.; Miller, M. Maximum h-colourable subgraph problem in balanced graphs. *Inf. Process. Lett.* **1998**, *65*, 301–303. [CrossRef]
3. Chang, M.-S.; Kloks, T.; Lee, C.-M. Maximum clique transversals. In Proceedings of the 27th International Workshop on Graph-Theoretic Concepts in Computer Science, Boltenhagen, Germany, 14–16 June 2001; Lecture Notes in Computer Science; Springer: Berlin/Heidelberg, Germany, 2001; Volume 2204, pp. 32–43.
4. Lee, C.-M. Variations of maximum-clique transversal sets on graphs. *Ann. Oper. Res.* **2010**, *181*, 21–66. [CrossRef]
5. Lee, C.-M.; Chang, M.-S. Signed and minus clique-transversal function on graphs. *Inf. Process. Lett.* **2009**, *109*, 414–417. [CrossRef]
6. Wang, H.; Kang, L.; Shan, E. Signed clique-transversal functions in graphs. *Int. J. Comput. Math.* **2010**, *87*, 2398–2407. [CrossRef]
7. Xu, G.; Shan, E.; Kang, L.; Chang, T.C.E. The algorithmic complexity of the minus clique-transversal problem. *Appl. Math. Comput.* **2007**, *189*, 1410–1418.
8. Chang, M.S.; Chen, Y.H.; Chang, G.J.; Yan, J.H. Algorithmic aspects of the generalized clique transversal problem on chordal graphs. *Discret. Appl. Math.* **1996**, *66*, 189–203.
9. Lee, C.-M.; Chang, M.-S. Variations of Y-dominating functions on graphs. *Discret. Math.* **2008**, *308*, 4185–4204. [CrossRef]
10. Balachandran, V.; Nagavamsi, P.; Rangan, C.P. Clique transversal and clique independence on comparability graphs. *Inf. Process. Lett.* **1996**, *58*, 181–184. [CrossRef]
11. Lee, C.-M.; Chang, M.-S. Distance-hereditary graphs are clique-perfect. *Discret. Appl. Math.* **2006**, *154*, 525–536. [CrossRef]
12. Bonomo, F.; Durán, G.; Lin, M.C.; Szwarcfiter, J.L. On balanced graphs. *Math. Program.* **2006**, *105*, 233–250. [CrossRef]
13. Lehel, J.; Tuza, Z. Neighborhood perfect graphs. *Discret. Math.* **1986**, *61*, 93–101. [CrossRef]
14. Chang, G.J.; Farber, M.; Tuza, Z. Algorithmic aspects of neighborhood numbers. *SIAM J. Discret. Math.* **1993**, *6*, 24–29. [CrossRef]
15. Fulkerson, D.R.; Hoffman, A.; Oppnheim, R. On balnaced matrices. *Math. Program. Study* **1974**, *1*, 120–132.
16. Argiroffo, G.; Leoni, V.; Torres, P. On the complexity of $\{k\}$-domination and k-tuple domination in graphs. *Inf. Process. Lett.* **2015**, *115*, 556–561. [CrossRef]
17. Faria, L.; Hon, W.-K.; Kloks, T.; Liu, H.-H.; Wang, T.-M.; Wang, Y.-L. On complexities of minus domination. *Discret. Optim.* **2016**, *22*, 6–19. [CrossRef]
18. Liao, C.-S.; Chang, G.J. k-tuple domination in graphs. *Inf. Process. Lett.* **2003**, *87*, 45–50. [CrossRef]
19. Brandstädt, A.; Le, V.B.; Spinrad, J.P. *Graph Classes–A Survey, SIAM Monographs on Discrete Math and Applications*; Society for Industrial and Applied Mathematics: Philadelphia, PA, USA, 1999.
20. Fulkerson, D.R.; Gross, O. Incidence matrices and interval graphs. *Pac. J. Math.* **1965**, *15*, 835–855. [CrossRef]

21. Brandstädt, A.; Dragan, F.F.; Chepoi, V.D.; Voloshin, V.I. Dually chordal graphs. *SIAM J. Discret. Math.* **1998**, *11*, 437–455. [CrossRef]
22. Dragan, F.F. HT-graphs: Centers, connected *r*-domination, and Steiner trees. *Comput. Sci. J. Mold.* **1993**, *1*, 64–83.
23. Moscarini, M. Doubly chordal graphs, Steiner trees, and connected domination. *Networks* **1993**, *23*, 59–69. [CrossRef]
24. Prisner, E.; Szwarcfiter, J.L. Recognizing clique graphs of directed and rooted path graphs. *Discret. Appl. Math.* **1999**, *94*, 321–328. [CrossRef]
25. Brandstädt, A.; Chepoi, V.D.; Dragan, F.F. Clique *r*-domination and clique *r*-packing problems on dually chordal graphs. *SIAM J. Discret. Math.* **1997**, *10*, 109–127. [CrossRef]
26. Rose, D.J. Triangulated graphs and the elimination process. *J. Math. Anal. Appl.* **1970**, *32*, 597–609. [CrossRef]
27. Patil, H.P. On the structure of *k*-trees. *J. Comb. Inf. Syst. Sci.* **1986**, *11*, 57–64.
28. Guruswami, V.; Rangan, C.P. Algorithmic aspects of clique-transversal and clique-independent sets. *Discret. Appl. Math.* **2000**, *100*, 183–202. [CrossRef]
29. Wood, D.R. On the maximum number of cliques in a graph. *Graphs Comb.* **2007**, *23*, 337–352. [CrossRef]
30. Uehara, R. *NP-Complete Problems on a 3-Connected Cubic Planar Graph and Their Applications*; Technical Report TWCU-M-0004; Tokyo Woman's Christian University: Tokyo, Japan, 1996.

MDPI

Article

A Quasi-Hole Detection Algorithm for Recognizing k-Distance-Hereditary Graphs, with $k < 2$

Serafino Cicerone

Department of Information Engineering, Computer Science and Mathematics, University of L'Aquila,
I-67100 L'Aquila, Italy; serafino.cicerone@univaq.it

Abstract: Cicerone and Di Stefano defined and studied the class of k-distance-hereditary graphs, i.e., graphs where the distance in each connected induced subgraph is at most k times the distance in the whole graph. The defined graphs represent a generalization of the well known distance-hereditary graphs, which actually correspond to 1-distance-hereditary graphs. In this paper we make a step forward in the study of these new graphs by providing characterizations for the class of all the k-distance-hereditary graphs such that $k < 2$. The new characterizations are given in terms of both forbidden subgraphs and cycle-chord properties. Such results also lead to devise a polynomial-time recognition algorithm for this kind of graph that, according to the provided characterizations, simply detects the presence of quasi-holes in any given graph.

Keywords: distance-hereditary graphs; stretch number; recognition problem; forbidden subgraphs; hole detection

check for
updates

Citation: Cicerone, S. A Quasi-Hole Detection Algorithm for Recognizing k-Distance-Hereditary Graphs, with $k < 2$. *Algorithms* **2021**, 14, 105. https://doi.org/10.3390/a14040105

Academic Editor: Frank Werner

Received: 10 February 2021
Accepted: 23 March 2021
Published: 25 March 2021

Publisher's Note: MDPI stays neutral with regard to jurisdictional claims in published maps and institutional affiliations.

1. Introduction

Distance-hereditary graphs have been introduced by Howorka [1], and are defined as those graphs in which every connected induced subgraph is isometric; that is, the distance between any two vertices in the subgraph is equal to the one in the whole graph. Therefore, any connected induced subgraph of any distance-hereditary graph G "inherits" its distance function from G. Formally:

Definition 1 (from [1]). *A graph G is a distance-hereditary graph if, for each connected induced subgraph G' of G, the following holds: $d_{G'}(x,y) = d_G(x,y)$, for each $x,y \in G'$.*

This kind of graph have been rediscovered many times (e.g., see [2]). Since their introduction, dozens of papers have been devoted to them, and different kinds of characterizations have been found: metric, forbidden subgraphs, cycle/chord conditions, level/neighborhood conditions, generative, and more (e.g., see [3]). Among such results, the generative properties resulted as the most fruitful for algorithmic applications, since they allowed researchers to efficiently solve many combinatorial problems in the class of distance-hereditary graphs (e.g., see [4–9]).

From an applicative point of view, distance-hereditary graphs are mainly attractive due to their basic metric property. For instance, these graphs can model unreliable communication networks [10,11] in which vertex failures may occur: at a given time, if sender and receiver are still connected, any message can be still delivered without increasing the length of the path used to reach the receiver.

Since in communication networks this property could be considered too restrictive, in [12] the class of k-distance-hereditary graphs has been introduced. These graphs can model unreliable networks in which messages can eventually reach the destination traversing a path whose length is at most k times the length of a shortest path computed in absence of vertex failures. The minimum k a network guarantees regardless the failed vertices is called *stretch number*. Formally:

Definition 2 (from [12]). *Given a real number $k \geq 1$, a graph G is a k-distance-hereditary graph if, for each connected induced subgraph G' of G, the following holds:* $d_{G'}(x,y) \leq k \cdot d_G(x,y)$, *for each $x,y \in G'$.*

The class of all the k-distance-hereditary graphs is denoted by $DH(k)$. Concerning this class of graphs, the following relationships hold:

- $DH(1)$ coincides with the class of distance-hereditary graphs;
- $DH(k_1) \subseteq DH(k_2)$, for each $k_1 \leq k_2$.

Additional results about the class hierarchy $DH(k)$ can be found in [13,14]. It is worth to notice that this hierarchy is *fully general*; that is, for each arbitrary graph G there exists a number k such that $G \in DH(k)$. It follows that the stretch number of G, denoted as $s(G)$, is the smallest number t such that G belongs to $DH(t)$. In [12], it has been shown that the stretch number $s(G)$ of any connected graph G can be computed as follows:

- the stretch number of any pair $\{u,v\}$ of distinct vertices is defined as $s_G(u,v) = D_G(u,v)/d_G(u,v)$, where $D_G(u,v)$ is the length of any longest induced path between u and v, and $d_G(u,v)$ is the distance between the same pair of vertices;
- $s(G) = \max_{\{u,v\}} s_G(u,v)$.

It follows that for any non-trivial graph G with $n \geq 4$ vertices, by simply maximizing $D(u,v)$ and minimizing $d(u,v)$, we get $s(G) \leq (n-2)/2$. From the above relationship about $s(G)$, we get that the stretch number is always a rational number. Interestingly, it has been shown that there are some rational numbers that cannot be stretch numbers. Formally, a positive rational number t is called *admissible stretch number* if there exists a graph G such that $s(G) = t$. The following result characterizes which numbers are admissible stretch numbers.

Theorem 1 (from [14]). *A rational number t is an admissible stretch number if and only if $t = 2 - \frac{1}{i}$, for some integer $i \geq 1$, or $t \geq 2$.*

Apart from the interesting general results found for the classes $DH(k)$, the original motivation was studying how (if possible) to extend the known algorithmic results from the base class, namely $DH(1)$, to $DH(k)$ for some constant $k > 1$. According to Theorem 1, in this work we are interested in studying the class containing each graph G such that $s(G) < 2$. Since this class contains graphs with stretch number *strictly* less than two, throughout this paper it will be denoted by $sDH(2)$.

Results. In this work, we provide three results for the class $sDH(2)$, namely two different characterizations and a recognition algorithm (notice that the characterizations have already been presented in [13] but with omitted proofs). The first characterization is based on listing all the minimal forbidden subgraphs for each graph in the class. It is interesting to observe the similarity with the corresponding result for the class $DH(1)$:

- **(adapted from [2])** $G \in DH(1)$ if and only if the following graphs are not induced subgraphs of G:

    - holes H_n, for each $n \geq 5$;
    - cycles C_5 with $cd(C_5) = 1$;
    - cycles C_6 with $cd(C_6) = 1$.

- **(this paper)** $G \in sDH(2)$ if and only if the following graphs are not induced subgraphs of G:

    - holes H_n, for each $n \geq 6$;
    - cycles C_6 with $cd(C_6) = 1$;
    - cycles C_7 with $cd(C_7) = 1$;
    - cycles C_8 with $cd(C_8) = 1$.

Here we used the notion of "chord distance" $cd(C)$ to express the position of possible chords within any cycle C (see Section 2 for a formal definition). Notice that in [14] a similar result has been provided for the generic class $DH(2 - \frac{1}{i})$, $i > 1$.

The second result is a characterization based on a cycle-chord property. As in the previous case, notice the similarity with the corresponding result for the class $DH(1)$:

- **(from [12])** $G \in DH(1)$ if and only if $cd(C_n) > 1$ for each cycle C_n, $n \geq 5$, of G;
- **(this paper)** $G \in sDH(2)$ if and only if $cd(C_n) > 1$ for each cycle C_n, $n \geq 6$, of G.

The last result is a recognition algorithm for graphs belonging to $sDH(2)$ that works in $O(n^2 m^2)$ time and $O(m^2)$ space. Basically, this algorithm exploits the result based on the cycle-chord property and, as a consequence, simply detects *quasi-holes* in any graph. A quasi-hole is any cycle with at least five vertices and chord-distance at most one (i.e., all the possible chords of the cycle must be incident to the same vertex). This algorithm is obtained by adapting the algorithm provided in [15] for detecting holes (i.e., any cycle with at least five vertices and no chords).

Outline. The paper is organized as follows. In Section 2, we introduce notation and basic concepts used throughout the paper. Sections 3 and 4 are devoted to providing the characterization based on minimal forbidden subgraphs and cycle-chord conditions for graphs in $sDH(2)$, respectively. In Section 5, we provide the algorithm for detecting quasi-holes and hence to solve the recognition problem for the class $sDH(2)$. Finally, Section 6 provides some concluding remarks.

2. Notation and Basic Concepts

We consider finite, simple, loop-less, undirected, and unweighted graphs $G = (V, E)$ with vertex set V and edge set E. A *subgraph* of G is a graph having all its vertices and edges in G. Given $S \subseteq V$, the *induced subgraph* $G[S]$ of G is the maximal subgraph of G with vertex set S. Given $u \in V$, $N_G(u)$ denotes the set of *neighbors* of u in G, and $N_G[u] = N_G(u) \cup \{u\}$.

A sequence of pairwise distinct vertices $(x_0, x_1, \ldots, x_k)$ is a *path* in G if $(x_i, x_{i+1}) \in E$ for $0 \leq i < k$; vertex x_i, for each $0 < i < k$, is an *internal vertex* of that path. A *chord* of a path is any edge joining two non-consecutive vertices in the path, and a path is an *induced path* if it has no chords. We denote by P_k any induced path with $k \geq 3$ vertices (e.g., an induced path on three vertices is denoted as P_3 whereas an induced path on four vertices is denoted as P_4). Two vertices x and y are *connected* in G if there exists a path $(x, \ldots, y)$ in G. A graph is *connected* if every pair of vertices is connected.

A *cycle* in G is a path $(x_0, x_1, \ldots, x_{k-1})$ where also $(x_0, x_{k-1}) \in E$. Two vertices x_i and x_j are *consecutive* in the cycle $(x_0, x_1, \ldots, x_{k-1})$ if $j = (i+1) \bmod k$ or $i = (j+1) \bmod k$. A *chord* of a cycle is an edge joining two non-consecutive vertices in the cycle. We denote by C_k any cycle with $k \geq 3$ vertices, whereas H_k denotes a *hole*, i.e., a cycle C_k, $k \geq 5$, without chords. The *chord distance* of a cycle C_k is denoted by $cd(C_k)$ and is defined as the minimum number of consecutive vertices in C_k such that every chord of C_k is incident to some of such vertices (see Figure 1 for an example of chord distance). We assume $cd(H_k) = 0$.

The length of any shortest path between two vertices x and y in a graph G is called *distance* and is denoted by $d_G(x, y)$. Moreover, the length of any longest induced path between them is denoted by $D_G(x, y)$. If x and y are distinct vertices, we use the symbols $p_G(x, y)$ and $P_G(x, y)$ to denote any shortest and any longest induced path between x and y, respectively. Sometimes, when no ambiguity occurs, we also use $p_G(x, y)$ and $P_G(x, y)$ to denote the sets of vertices belonging to the corresponding paths. If $d_G(x, y) \geq 2$, then $\{x, y\}$ is a *cycle-pair* if there exist two induced paths $p_G(x, y)$ and $P_G(x, y)$ such that $p_G(x, y) \cap P_G(x, y) = \{x, y\}$. In other words, if $\{x, y\}$ is a cycle-pair, then there exist induced paths $p_G(x, y)$ and $P_G(x, y)$ such that the vertices in $p_G(x, y) \cup P_G(x, y)$ form a cycle in G; this cycle is denoted by $G[x, y]$. In Figure 1 $\{v_3, v_6\}$ is a cycle-pair that induces the cycle $(v_3, v_4, v_5, v_6, v_1)$; in particular, $G[v_3, v_6]$ is induced by $p_G(v_3, v_6) = (v_3, v_1, v_6)$ and $P_G(v_3, v_6) = (v_3, v_4, v_5, v_6)$. We use the symbol $\mathcal{S}(G)$ to denote the set containing all pairs $\{u, v\}$ of connected vertices that induce the stretch number of G, namely $\mathcal{S}(G) =$

$\{\{x,y\} : s_G(x,y) = s(G)\}$. The following lemma states that cycle-pairs are useful to determine the stretch number.

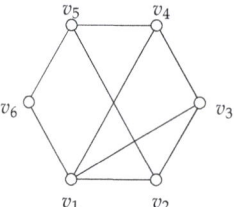

Figure 1. The chord distance of this C_6 graph is two because: (i) vertices v_1 and v_2 are consecutive in the cycle, (ii) every chord is incident to one of such vertices, and (iii) there is no other set with less than two vertices with the same properties.

Lemma 1 (from [12]). *Let G be a graph such that $s(G) > 1$. The following relationships hold:*
(i) $d_G(u,v) \geq 2$ for each pair $\{u,v\}$ such that $\{u,v\} \in S(G)$,
(ii) there exists a cycle-pair $\{u,v\}$ that induces the stretch number of G, that is $\{u,v\} \in S(G)$.

This lemma suggests that studying $s(G)$ concerns the analysis of cycles in G. In particular, if $\{u,v\}$ is a cycle-pair that belongs to $S(G)$, then the cycle $G[u,v]$ is called *inducing-stretch cycle* for G. In Figure 1, the represented graph G belongs to $DH(3/2)$; moreover, both $\{v_3, v_5\}$ and $\{v_3, v_6\}$ are cycle-pairs in $S(G)$, and $(v_1, v_3, v_4, v_5, v_6)$ is the corresponding inducing-stretch cycle.

3. A Characterization Based on Forbidden Subgraphs

A well known characterization based on *minimal forbidden subgraphs* has been provided for the class of distance-hereditary graphs.

Theorem 2 (from [2]). *A graph G is a distance-hereditary graph if and only if it does not contain, as an induced subgraph, any of the following graphs: the hole H_n, $n \geq 5$, the house, the fan, and the domino (cf. Figure 2).*

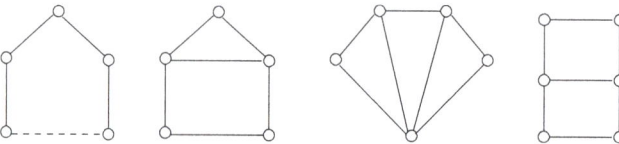

Figure 2. The minimal forbidden subgraphs of distance-hereditary graphs: from left to right, the *hole*, the *house*, the *fan*, and the *domino*. Dashed lines represent paths of length at least one.

This result can be easily reformulated, and simplified, by using the notion of chord distance. In particular, it is possible to characterize in a compact way all the forbidden subgraphs by using just the notion of chord distance as follows:

- G is a distance-hereditary graph if and only if the following graphs are not induced subgraphs of G:

  (i) H_n, for each $n \geq 5$;
  (ii)   cycles C_5 with $cd(C_5) = 1$;
  (iii) cycles C_6 with $cd(C_6) = 1$.

It is worth to notice that in this way we do not consider the minimal subgraphs only (cf. Figure 3).

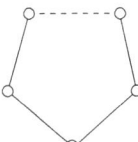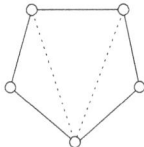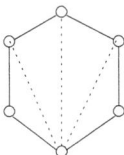

Figure 3. The forbidden subgraphs of DH(1) expressed according to the notion of chord distance. Dashed lines represent paths of length at least one. Dotted lines represent chords that may or may not exist.

In the following we provide a characterization similar to that of Theorem 2 for any graph $G \in sDH(2)$. Before giving such a result, we need to recall the following technical lemma.

Lemma 2. *Let G be a graph and let $G[x, y]$ be an inducing-stretch cycle of G defined by the induced paths $P_G(x, y) = (x, u_1, u_2, \ldots, u_{p-1}, y)$ and $p_G(x, y) = (x, v_1, v_2, \ldots, v_{q-1}, y)$. If $d(x, y) \geq 3$ then v_1 must be incident to chords of the cycle $G[x, y]$.*

Proof. Since $G[x, y]$ is an inducing-stretch cycle of G, then $s(G) = \frac{p}{q}$. If v_1 is not incident to any chords of $G[x, y]$, then the induced paths $P_G(v_1, y) = (v_1, x, u_1, u_2, \ldots, u_{p-1}, y)$ and $p_G(v_1, y) = (v_1, v_2, \ldots, v_{q-1}, y)$ imply $s_G(v_1, y) = \frac{p+1}{q-1} > \frac{p}{q}$, a contradiction. □

Let G be any graph. According to Lemma 1, let us consider an inducing-stretch cycle $G[x, y]$ of G. Assume that $G[x, y]$ is formed by the vertices of the induced paths $P_G(x, y) = (x, u_1, u_2, \ldots, u_{p-1}, y)$ and $p_G(x, y) = (x, v_1, v_2, \ldots, v_{q-1}, y)$. Since $P_G(x, y)$ and $p_G(x, y)$ are induced paths, each chord of $G[x, y]$ (if any) joins vertices v_i and u_j, with $1 \leq i \leq q - 1$ and $1 \leq j \leq p - 1$. When some vertex v_i is incident to chords of $G[x, y]$, we denote by (v_i, u_{l_i}) and (v_i, u_{r_i}) the *leftmost* and *rightmost* chords of v_i, respectively. Formally, the indices l_i and r_i are defined as follows:

- $l_i = \min\{i' \mid 1 \leq i' \leq p - 1 \text{ and } (v_i, u_{i'}) \text{ is a chord of } G[x, y]\}$
- $r_i = \max\{i' \mid 1 \leq i' \leq p - 1 \text{ and } (v_i, u_{i'}) \text{ is a chord of } G[x, y]\}$

Theorem 3. *Let G be a graph. $G \in sDH(2)$ if and only if the following graphs are not induced subgraphs of G:*

(i) H_n, for each $n \geq 6$;
(ii) cycles C_6 with $cd(C_6) = 1$;
(iii) cycles C_7 with $cd(C_7) = 1$;
(iv) cycles C_8 with $cd(C_8) = 1$.

Proof. ($\Rightarrow$) Each provided hole and cycle has stretch number greater or equal to 2, and hence it cannot be an induced subgraph of G.

($\Leftarrow$) We prove that if $s(G) \geq 2$, then G contains one of the subgraphs in items (i), (ii), (iii), or (iv), or G contains a proper induced subgraph G' such that $s(G') \geq 2$. In the latter case, we can recursively apply to G' the following proof.

According to Lemma 1, consider an inducing-stretch cycle $G[x, y]$ of G and assume it is formed by the vertices of the induced paths $P_G(x, y) = (x, u_1, u_2, \ldots, u_{p-1}, y)$ and $p_G(x, y) = (x, v_1, v_2, \ldots, v_{q-1}, y)$. Notice that, since $P_G(x, y)$ and $p_G(x, y)$ are induced paths, each possible chord of $G[x, y]$ joins vertices v_i and u_j, with $1 \leq i \leq q - 1$ and $1 \leq j \leq p - 1$.

Since $\frac{p}{q} \geq 2$ by hypotheses, then $q \geq 2$ by Item (i) of Lemma 1, and hence $p \geq 4$. According to the value of q, we analyze two different cases:

$q = 2$: In this case, if $G[x, y]$ is chordless, then it corresponds to a hole as described in Item (i). If the chord distance of $G[x, y]$ is equal to 1, all chords are incident to v_1. According to p, we have:

$p = 4$: $G[x, y]$ corresponds to the cycle in Item (ii);

$p = 5$: $G[x, y]$ corresponds to the cycle in Item (iii);

$p = 6$: $G[x, y]$ corresponds to the cycle in Item (iv);

$p \geq 7$: Let (v_1, u_{l_1}) be the leftmost chord of v_1. If $l_1 \geq 4$ the cycle $(v_1, x, u_1, u_2, \ldots, u_{l_1})$ corresponds to the cycle in Item (i). When $l_1 \leq 3$, consider the subgraph G' induced by the vertices in the cycle $(v_1, u_{l_1}, u_{l_1+1}, \ldots, u_{p-1}, y)$. The induced paths $P' = (u_{l_1}, u_{l_1+1}, \ldots, u_{p-1}, y)$ and $p' = (u_{l_1}, v_1, y)$ provide the following lower bound for $s_{G'}$:

$$s_{G'}(u_{l_1}, y) \geq \frac{p-l_1}{2} \geq \frac{7-3}{2} = 2.$$

Hence, G' is a proper subgraph of G with $s(G') \geq 2$. The statement follows by recursively applying to G' this proof.

$q \geq 3$: In this case, according to Lemma 2, v_1 must be incident to chords. We now analyze two cases with respect to the value of r_1, (v_1, u_{r_1}) being the rightmost chord of v_1:

$r_1 \geq 4$: Consider the subgraph G'' induced by the vertices in the cycle $(v_1, x, u_1, u_2, \ldots, u_{r_1})$. In this case, the induced paths $P'' = (x, u_1, u_2, \ldots, u_{r_1})$ and $p'' = (x, v_1, u_{r_1})$ provide the following lower bound for $s_{G''}$: $s_{G''}(x, u_{r_1}) \geq r_1/2 \geq 2$. Hence, G'' is a proper subgraph of G with $s(G'') \geq 2$. The statement follows by recursively applying to G'' this proof.

$r_1 \leq 3$: in this case the induced paths $P''' = (v_1, u_{r_1}, u_{r_1+1}, \ldots, u_{p-1}, y)$ and $p''' = (v_1, v_2, \ldots, v_{q-1}, y)$ provide the following lower bound for $s_G(v_1, y)$:

$$s_G(v_1, y) \geq \frac{p-2}{q-1}.$$

Since $\frac{p-2}{q-1} \geq \frac{p}{q}$ is equivalent to $\frac{p}{q} \geq 2$ (which holds by hypothesis), then the subgraph G''' induced by the vertices in both P''' and p''' is a proper subgraph of G with stretch $p^*/q^* \geq 2$ and $q^* = q - 1$. Hence, the statement follows by recursively applying to G''' this proof.

This concludes the proof. $\square$

Figures 3 and 4 summarize the characterizations based on forbidden subgraphs for classes DH(1) and sDH(2), respectively. Figure 5 provides the list of all the *minimal* forbidden subgraphs of any graph in sDH(2).

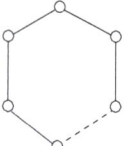 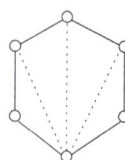 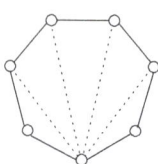 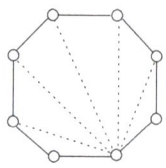

Figure 4. The forbidden subgraphs of graphs having stretch number less than 2. Dashed (dotted, respectively) lines represent paths of length at least one (chords that may or may not exist, respectively).

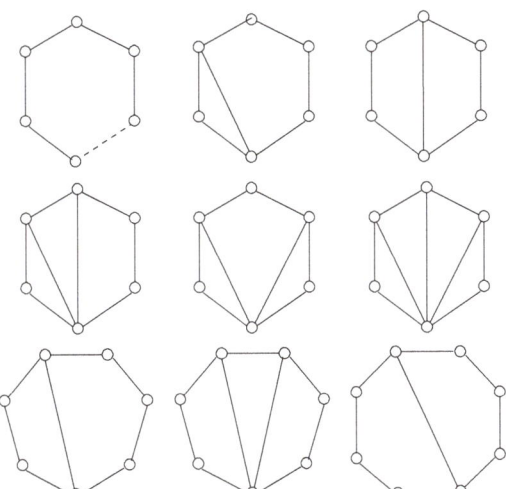

Figure 5. The *minimal* forbidden subgraphs of any graph with stretch number less than 2. Dashed lines represent paths of length at least one.

4. A Characterization Based on Cycle-Chord Conditions

For the class of distance-hereditary graphs, Howorka provided the following well known characterization based on cycle-chord conditions.

Theorem 4 (from [1]). *Let G be a graph. $G \in \mathrm{DH}(1)$ if and only if each cycle C_n, $n \geq 5$, of G has two crossing chords.*

In [12], this result has been reformulated in terms of chord distance:

Theorem 5 (from [12]). *Let G be a graph. $G \in \mathrm{DH}(1)$ if and only if $cd(C_n) > 1$ for each cycle C_n, $n \geq 5$, of G.*

In the remainder of this section, we provide a similar characterization for graphs belonging to $s\mathrm{DH}(2)$.

Lemma 3. *Let G be a graph. If $s(G) = 2$ then G contains, as induced subgraph, a cycle C_6 with chord distance at most 1.*

Proof. According to Lemma 1, consider an inducing-stretch cycle $G[x,y]$ of G. Since $s(G) = 2$, assume that $G[x,y]$ is formed by the vertices of the induced paths $P_G(x,y) = (x, u_1, u_2, \ldots, u_{2s-1}, y)$ and $p_G(x,y) = (x, v_1, v_2, \ldots, v_{s-1}, y)$, with $s \geq 2$.

If $s = 2$ then the proof is concluded. In fact, cycle $G[x,y]$ has 6 vertices and every chord of $G[x,y]$ (if any) is incident to v_1.

In the remainder of the proof assume $s \geq 3$. In this case, according to Lemma 2, v_1 is incident to chords of $G[x,y]$. Let (v_1, u_{r_1}) be the rightmost chord incident to v_1. We analyze different cases according to the value of r_1.

- Assume $r_1 > 4$. In this case, the induced paths $(x, u_1, u_2, \ldots, u_{r_1})$ and (x, v_1, u_{r_1}) provide a stretch number $s_G(x, u_{r_1}) \geq \frac{r_1}{2} > 2$, a contradiction.
- Assume $r_1 \leq 2$. In this case, the induced paths $(v_1, u_{r_1}, u_{r_1+1}, \ldots, u_{2s-1}, y)$ and $(v_1, v_2, \ldots, v_{s-1}, y)$ provide the following lower bound on $s_G(v_1, y)$:

$$s_G(v_1, y) \geq \frac{2s - r_1 + 1}{s - 1} \geq \frac{2s - 2 + 1}{s - 1} = 2 + \frac{1}{s - 1}.$$

This contradicts $s(G) = 2$.

It follows that either $r_1 = 4$ or $r_1 = 3$. In the first case the cycle $(v_1, x, u_1, u_2, u_3, u_4)$ represents the requested cycle C_6: chords of $G[x, y]$ (if any) are all incident to v_1. In the second case consider the induced paths $(v_1, u_{r_1}, u_{r_1+1}, \ldots, u_{2s-1}, y)$ and $(v_1, v_2, \ldots, v_{s-1}, y)$. These paths induce the following lower bound on $s_G(v_1, y)$:

$$s_G(v_1, y) \geq \frac{2s - r_1 + 1}{s - 1} = \frac{2s - 3 + 1}{s - 1} = 2.$$

Hence, the above paths induce a proper subgraph G' of G with stretch number 2. Hence, this proof can be recursively applied to G'. □

Lemma 4. *Let G be a graph. $s(G) \geq 2$ if and only if G contains, as an induced subgraph, a cycle C_n, $n \geq 6$, with chord distance at most 1.*

Proof. ($\Leftarrow$) Trivial.
($\Rightarrow$) If $s(G) = 2$, then it is sufficient to use Lemma 3. Now, let us assume that $s(G) = p/q > 2$ such that p and q are coprime. By Lemma 1, if $G[x, y]$ is an inducing-stretch cycle of G, according to the hypotheses, we may assume that $G[x, y]$ is formed by the vertices of the induced paths $P_G(x, y) = (x, u_1, u_2, \ldots, u_{p \cdot s - 1}, y)$ and $p_G(x, y) = (x, v_1, v_2, \ldots, v_{q \cdot s - 1}, y)$, with $s \geq 1$.
If $d(x, y) = 2$, then $G[x, y]$ contains at least 6 vertices and all its chords (if any) are incident to v_1. Then, $G[x, y]$ corresponds to the requested cycle.
In the remainder, assume that $d(x, y) \geq 3$. In this case, by Lemma 2, vertex v_1 is incident to chords of $G[x, y]$: let (v_1, u_{r_1}) be the rightmost chord incident to it.
If $r_1 \leq 3$, then the two induced paths $(v_1, u_{r_1}, u_{r_1+1}, \ldots, u_{p \cdot s - 1}, y)$ and $(v_1, v_2, \ldots, v_{q \cdot s - 1}, y)$ provide the following lower bound for $s_G(v_1, y)$:

$$s_G(v_1, y) \geq \frac{p \cdot s - r_1 + 1}{q \cdot s - 1}.$$

Now we show that

$$\frac{p \cdot s - r_1 + 1}{q \cdot s - 1} > \frac{p}{q}. \tag{1}$$

It can be easily observed that Equation (1) is equivalent to

$$\frac{p}{q} > r_1 - 1. \tag{2}$$

Since $r_1 \leq 3$ and $p/q > 2$ by hypothesis, then Equation (2) holds. This implies that $s_G(v_1, y) > p/q$, a contradiction.
Then, it follows that $r_1 \geq 4$. In this case, $C = (x, u_1, u_2, \ldots, u_{r_1}, v_1)$ is an induced cycle with $r_1 + 2 \geq 6$ vertices and chord distance at most 1 (In C, all the possible chords are incident to v_1). This concludes the proof. □

This lemma can be reformulated so that it directly provides a characterization for the graphs under consideration.

Theorem 6. *Let G be a graph. $G \in \text{sDH}(2)$ if and only if $cd(C_n) > 1$ for each cycle C_n, $n \geq 6$, of G.*

Compare Theorems 5 and 6 to observe the similarity between the cycle-chord characterizations of graphs with stretch number equal to 1 and graphs with stretch number less than 2, respectively.

5. Recognition Algorithm

The distance-hereditary graphs, i.e., graphs in DH(1), can be recognized in linear time [16], while the recognition problem for the generic class DH(k), k not fixed, is co-NP-complete [12]. For small and fixed values of k, in [14] a partial answer to this basic problem is given. In particular, Lemma 1 states that for $k < 2$, only specific rational numbers may act as stretch numbers. In [14], a characterization for each class DH($2 - 1/i$), $i > 1$, has been provided, and such a characterization led to a polynomial time algorithm for the recognition problem for the class DH($2 - 1/i$), with fixed $i > 1$. Unfortunately, the running time of this algorithm is bounded by $O(n^{3i+2})$.

In this section, we propose a polynomial-time algorithm for solving the recognition problem for the class sDH(2) according to the following approach. Lemma 4 provides a characterization for all graphs not belonging to sDH(2). It is based on detecting whether a given graph G contains or not an induced cycle C_n, $n \geq 6$, with chord distance at most 1. Now, assume that we have an algorithm $\mathcal{A}$ returning *true* if and only if a given graph G contains such a cycle. Then, to recognize whether $G \in s$DH(2) we can simply use $\mathcal{A}$ on G and certify the membership if and only if $\mathcal{A}$ return *false*. In the remainder of this section we show that such an algorithm $\mathcal{A}$ can be defined.

5.1. An Existing Hole Detection Algorithm

We remind that H_k denotes a hole, i.e., a chordless cycle with $k \geq 5$ vertices. In [15], Nikolopoulos and Palios provided the following result about the hole detection problem.

Theorem 7 (from [15]). *Given any connected graph $G = (V, E)$ with $|V| = n$ and $|E| = m$, it is possible to determine whether G contains a hole in $O(m^2)$ time and $O(nm)$ space.*

They also extended their result to larger versions of holes.

Corollary 1 (from [15]). *Let $G = (V, E)$ be a connected graph with $|V| = n$ and $|E| = m$, and let $k \geq 5$ be a constant. It is possible to determine whether G contains a hole on at least k vertices in $O(nm^{p-1})$ time and $O(m^{p-1})$ space if $k = 2p$, and in $O(n + m^p)$ time and $O(nm^{p-1})$ space if $k = 2p + 1$.*

Therefore, according to this corollary, it is possible to check whether G contains a hole H_k, with $k \geq 6$ vertices, in $O(nm^2)$ time and $O(m^2)$ space.

5.2. Quasi-Hole Detection Algorithm

We call *quasi-hole* any cycle C_k such that $k \geq 5$ and $cd(C_k) \leq 1$. In what follows, we show that the hole-detection algorithms recalled in Theorem 7 and Corollary 1 can be adapted to detect quasi-holes in any connected graph G. This adapted version is called `QuasiHoleDetection` and it is described in pseudo-code as shown in Algorithms 1 and 2. The strategy behind `QuasiHoleDetection` is based on the following result:

Lemma 5. *A connected graph G contains a quasi-hole if and only if there exists a cycle $(v_0, v_1, \ldots, v_\ell)$, $\ell \geq 4$, in G such that each path $(v_i, v_{i+1}, v_{i+2}, v_{i+3})$, $i = 1, \ldots, \ell - 3$, is a P_4 of G.*

Proof. ($\Rightarrow$) If G contains a quasi-hole C_k then the vertices of C_k form a cycle fulfilling the conditions of the statement (where v_0 is the only vertex incident to possible chords of the cycle).

($\Leftarrow$) Suppose that G admits cycles as described in the statement, and let $C = (v_0, v_1, \ldots, v_\ell)$ be *the shortest* among such cycles. We now show that (*i*) C has at least 5 vertices and (*ii*) $cd(C) \leq 1$:

(*i*)   Since C fulfills the conditions of the statement, then C contains at least 5 vertices;

(*ii*) Suppose by contradiction that $cd(C) > 1$. Then, there must exist chords (v_i, v_j) with both v_i and v_j different from v_0. To each chord (v_i, v_j) not incident on v_0, we associate a "length" defined as $length(v_i, v_j) = |j - i|$. Now, let (v_l, v_r), with $l < r$, be a chord with minimum length. By definition, $0 < l < r \leq \ell$ holds. Since $(v_l, v_{l+1}, v_{l+2}, v_{l+3})$ is a P_4, then $r \geq l + 4$, and hence $C' = (v_l, v_{l+1}, \ldots, v_r)$ results to be a cycle with at least 5 vertices. Moreover, between v_i and v_j, for each $l \leq i < i + 2 \leq j \leq r$, $(i, j) \neq (l, r)$, cannot exist an edge, otherwise it would be a chord with length smaller than $length(v_l, v_r)$.

Since C' is a cycle with at least 5 vertices and with chord distance zero, then it contradicts the fact that C is the shortest among the cycles fulfilling the conditions of the statement. Hence, $cd(C) \leq 1$.

Since both the properties at points (i) and (ii) hold, it follows that C is a quasi-hole. □

Algorithm 1: A quasi-hole detection algorithm.

Algorithm: `QuasiHoleDetection`
Input : a connected undirected graph $G = (V, E)$
Output: "true" if G contains a quasi-hole, "false" otherwise.

1   calculate the adjacency matrix $M[]$ of G ;
2   **foreach** $v_1 \in V$ **do**
3   set each entry of the arrays $walked\_P_3[]$ and $AP[]$ to 0;
4      $base \leftarrow v_1$;
5      $AP[v_1] \leftarrow 1$;
6   **foreach** $(v_2, v_3) \in E$ **do**
7   **if** $(v_1, v_2) \in E$ *and* $v_1 \neq v_3$ **then**
8            $AP[v_2] \leftarrow 1$;
9   `Visit`$(base, v_1, v_2, v_3)$;
10           $AP[v_2] \leftarrow 0$;
11   **end**
12   **if** $(v_1, v_3) \in E$ *and* $v_1 \neq v_2$ **then**
13            $AP[v_3] \leftarrow 1$;
14   `Visit`$(base, v_1, v_3, v_2)$;
15           $AP[v_3] \leftarrow 0$;
16   **end**
17   **end**
18      $AP[v_1] \leftarrow 0$;
19 **end**
20 print "false".

The above lemma is used by the provided algorithm for the detection of quasi-holes in G. To this end, we associate to G a directed graph G^+ defined as follows:

- $\{v_{abc} \mid (a, b, c) \text{ is a } P_3 \text{ in the graph } G\}$ is the vertex set of G^+;
- $\{(v_{abc}, v_{bcd}) \mid (a, b, c, d) \text{ is a } P_4 \text{ in the graph } G\}$ is the edge set of G^+.

If (a, b, c) is a path P_3 of G, then both the vertices v_{abc} and v_{cba} belong to G^+. In a similar way, if (a, b, c, d) is a path P_4 of G, then the edges (v_{abc}, v_{bcd}) and (v_{dcb}, v_{cba}) must be contained in G^+. Hence, visiting G^+ is equivalent to proceeding along P_4s of G. It follows that the conditions of Lemma 5 on G can be verified by performing a revised DFS on G^+ (cf. [17]). In turn, the following lemma holds:

Lemma 6. *Let G be any connected graph, and let G^+ be its associated directed graph. By performing a DFS on G^+, if the DFS-path is $v_{u_0 u_1 u_2}, v_{u_1 u_2 u_3}, \ldots, v_{u_{k-2} u_{k-1} u_k}$, where $u_i \neq u_j$ for each $0 \leq i < j < k$ and $u_k = u_\ell$ for some ℓ such that $0 \leq \ell < k$, then $u_\ell, u_{\ell+1}, \ldots, u_{k-1}$ are vertices forming a cycle in G that fulfill Lemma 5. Conversely, if G contains a quasi-hole, the DFS on G^+ will meet a sequence of vertices in G^+ whose corresponding P_3s in G produce a path as the path $(v_1, v_2, \ldots, v_\ell)$ in the cycle as in Lemma 5.*

Algorithm 2: A recursive procedure used by `QuasiHoleDetection` to perform an adapted DFS.

Procedure: procedure `Visit`
Input : four vertices *base*, u_1, u_2, and u_3 of G

1 $AP[u_3] \leftarrow 1$;
2 $walked\_P_3[(u_1, u_2), u_3] \leftarrow 1$;
3 **foreach** $(u_3, u_4) \in E \setminus \{(u_3, u_2)\}$ **do**
4   **if** $u_4 = base$ **then**
5   **if** $AP.size \geq 5$ **then**
            // the active path determines a quasi-hole
6   print "true" ;
7   exit;
8   **else**
9   break
10   **end**
11   **else**
12   **if** $(u_2, u_4) \notin E$ *and* $(u_1 = base$ *or* $(u_1, u_4) \notin E)$ **then**
            // here, when $u_1 \neq base$, (u_1, u_2, u_3, u_4) forms a P_4 in G
13   **if** $AP[u_4] = 1$ **then**
               // the active path determines a hole
14   print "true" ;
15   exit;
16   **end**
17   **if** $walked\_P_3[(u_2, u_3), u_4] = 0$ **then**
18   `Visit`$(base, u_2, u_3, u_4)$;
19   **end**
20   **end**
21   **end**
22 **end**
23 $AP[u_3] \leftarrow 0$;

By following the same strategy used in [15], to reduce the space complexity required by G^+, the DFS on G^+ is simulated by performing a revised DFS directly on G. This revised DFS on G is implemented by Algorithm `QuasiHoleDetection` (cf. Figure 1).

At Line 1, the algorithm computes the adjacency matrix $M[]$ of G from its adjacency-list (we assume that G is provided as input according to this representation). $M[]$ is used to check the adjacency in constant time. At Line 2, each vertex v_1 of G is checked against the following possible role: v_1 belongs to a quasi-hole C and all the chords of C, if any, are adjacent to v_1. To perform this check, at Line 6 we consider each edge (v_2, v_3) in G: if this edge, along with (v_1, v_2) (cf. Line 7) or (v_1, v_3) (cf. Line 12), form a path with three vertices, then the algorithm tries to extend this path into the requested cycle by recursively calling the Procedure `Visit` (see Algorithm 2).

`Visit` works according to Lemma 5: in any step, it attempts to extend a path P_3 defined by (u_1, u_2, u_3) into P_4s of the form (u_1, u_2, u_3, u_4); then, for each such P_4, the procedure proceeds by extending the P_3 formed by (u_2, u_3, u_4) into P_4s of the form (u_2, u_3, u_4, u_5), and so on. In this situation, the *active-path* is first extended from (u_1, u_2, u_3) to (u_1, u_2, u_3, u_4), then to $(u_1, u_2, u_3, u_4, u_5)$ and so on. In case of backtracking, the last vertex is removed of the current active-path. By proceeding in this way, two cases may occur:

- the initial vertex v_1 (called *base* in the algorithm) is added again to the active-path (cf. Line 4). If the length of the active-path is 5 or more (cf. Line 5), then the graph contains a cycle fulfilling the conditions of Lemma 5 and hence a quasi-hole is found;

- at the end of the active-path there is a vertex different from *base* but already inserted in the active-path (cf. Lines 12–13). In this case, again the conditions of Lemma 5 apply, but now we are sure that a hole is found.

It is worth to remark that the ongoing active-path P on G and the ongoing DFS-path P^+ on G^+ contain exactly the same vertices: the elements of P correspond to the vertices of the P_3s associated with the elements of P^+ (in P, the repeated vertices of G in adjacent P_3s are present only once).

We now explain the role of the additional data structures $AP[\cdot]$ and $walked\_P_3[(\cdot,\cdot),\cdot]$. The former is an auxiliary array of size n used to check if a vertex appears in the "active path" computed so far; given u, $AP[u]$ is equal to 1 if u appears in the active path, 0 otherwise. Concerning the latter, during the visit on G^+, vertices that correspond to path P_3s of G are recorded so that they are not "visited" again. The entry $walked\_P_3[(u_1, u_2), u_3]$ equals one if and only if the vertices u_1, u_2, u_3 induce (u_1, u_2, u_3) as a path P_3 of G already encountered during the DFS, otherwise it equals zero. Since $walked\_P_3[(\cdot,\cdot),\cdot]$ has entries $walked\_P_3[(u_1, u_2), u_3]$ and $walked\_P_3[(u_2, u_1), u_3]$ for each edge $(u_1, u_2) \in E$ and for each $u_3 \in V$, then its size is $2m \cdot n$. Notice that Visit registers the entry of $walked\_P_3[]$ at the beginning, thus avoiding another execution on the same path P_3. In this way, Visit() is executed exactly once for each path P_3 of G.

Notice that the description of Visit() assures that starting from a P_3 formed by (u_1, u_2, u_3) we proceed to a P_3 formed by (u_2, u_3, u_4) only if (u_1, u_2, u_3, u_4) is a path P_4 of G. The only exception is when u_1 coincides with the starting vertex v_1 selected at Line 2 by QuasiHoleDetection: in such a case (u_1, u_2, u_3, u_4) may have chords from u_1. For this purpose, the initial vertex v_1 is assigned to the variable *base* (cf. Line 4 of the main algorithm) and it is later passed to Visit (cf. Lines 9 and 14 of the main algorithm).

We can now provide the following statement:

Theorem 8. *Given any connected graph $G = (V, E)$ with $|V| = n$ and $|E| = m$, it is possible to determine whether G contains a quasi-hole in $O(nm^2)$ time and $O(nm)$ space.*

Proof. According to the above description of QuasiHoleDetection, its correctness follows from Lemmas 5 and 6, and from the inherent execution of DFS on G^+. In the remainder of the proof we analyze the complexity of the algorithm about the required time and space.

As G is a connected graph, we get $n = O(m)$. Concerning the data structures used by the algorithm, we assume that from any edge (v_1, v_2) it is possible to access in constant time both its endpoints; alike, from any entry in the adjacency matrix $M[]$ of G corresponding to v_1 and v_2 it is possible to access in constant time the edge (v_1, v_2).

Consider first the time complexity of performing the revised DFS of G. The visit starts at Line 6, and proceeds by recursive calls to Visit. This recursive procedure checks each path (u_1, u_2, u_3) of G which is a P_3 and tries to extend it into a P_4 of the form (u_1, u_2, u_3, u_4). Notice that each set of vertices $\{u_1, u_2, u_3, u_4\}$ where (u_1, u_2, u_3) is a P_3 and u_4 is adjacent to u_3 is uniquely characterized by the ordered pair $((u_1, u_2), (u_3, u_4))$ where (u_1, u_2) and (u_3, u_4) are ordered pairs of adjacent vertices in G. Hence, the time required to perform the whole visit according to the recursive executions of Visit is $O(m^2)$. We can now determine the time complexity of QuasiHoleDetection. Step at Line 1 clearly takes $O(n^2)$ time. The subsequent loop at Line 2 is repeated $O(n)$ times, and for each step the algorithm requires $O(nm)$ time for the initialization at Line 3 and, as described before, $O(m^2)$ time for visiting G according to the recursive calls to Visit.

It follows that the final time complexity is $O(nm^2)$. The algorithm requires $O(nm)$ space: $O(n)$ and $O(nm)$ for the arrays $AP[]$ and $walked\_P_3[]$, respectively, and $O(n^2)$ for the adjacency matrix $M[]$ and the adjacency-list used to represent G. $\square$

5.3. Detecting Quasi-Hole on at Least k Vertices

As in [15], the strategy described above to define a quasi-hole detection algorithm can be generalized to built algorithms for the detection of quasi-holes on at least k vertices, with $k \geq 5$. For any input graph G, we consider the following family of directed graphs $G^{(t)}$:

- $\{v_{u_1 u_2 \cdots u_{t-1}} \mid (u_1, u_2, \ldots, u_{t-1})$ *is an induced path* P_{t-1} *in* $G\}$ is the vertex set of $G^{(t)}$,
- $\{(v_{u_1 u_2 \cdots u_{t-1}}, v_{u_2 u_3 \cdots u_t}) \mid (u_1, u_2, \ldots, u_t)$ *is an induced path* P_t *in* $G\}$ is the edge set of $G^{(t)}$.

By definition, $G \equiv G^{(2)}$ and $G^+ \equiv G^{(4)}$ where G^+ is the direct graph associated to G in Section 5.2. Therefore, in the same way that running DFS on $G^+ \equiv G^{(4)}$ allowed us to detect quasi-holes (on at least five vertices), running DFS on $G^{(k-1)}$ allows us to detect (extended) quasi-holes on at least k vertices, for each constant $k \geq 5$. This is ensured by the following statement, which represents a generalization of Lemma 5:

Lemma 7. *Given a constant $k \geq 5$, a graph G contains a quasi-hole on at least k vertices if and only if G contains a cycle $(u_0, u_1, \ldots, u_t)$, with $t \geq k - 1$, such that $(u_i, u_{i+1}, \ldots, u_{i+k-2})$ is an induced path P_{k-1} of G for each $i = 1, 2, \ldots, t - k + 2$.*

Lemmas 6 and 7 induce the following statement:

Corollary 2. *Let G be a connected graph and let $k \geq 5$ be a constant. Assume that a DFS is executed on $G^{(k-1)}$, the directed graph associated to G. If the active path computed by the DFS is*
$$v_{u_0 u_1 \cdots u_{k-3}}, v_{u_1 u_2 \cdots u_{k-2}}, \ldots, v_{u_{r-k+3} u_{r-k+4} \cdots u_r},$$
where $u_i \neq u_j$ for all $0 \leq i < j < r$, and $u_r = u_p$ for some p such that $0 \leq p < r$, then $u_p, u_{p+1}, \ldots, u_{r-1}$ are vertices forming a cycle in G that fulfill the conditions of Lemma 7. Conversely, if G contains a quasi-hole on at least k vertices, the DFS on $G^{(k-1)}$ will meet a sequence of vertices whose associated $P_{k-2}s$ in G form a path as the path $(u_1, u_2, \ldots, u_t)$ in the cycle of Lemma 7.

Additionally, in this situation we do not build $G^{(k-1)}$ since we implicitly run DFS on this associated graph. In particular, we process each unvisited P_{k-2} of G as follows: we try to extend the induced path P_{k-2} formed by $(u_0, u_1, \ldots, u_{k-3})$ into $P_{k-1}s$ of the form $(u_0, u_1, \ldots, u_{k-3}, u_{k-2})$; then, for each such P_{k-1}, we proceed by extending the P_{k-2} $(u_1, u_2, \ldots, u_{k-2})$ into $P_{k-1}s$, and so on. Since there exist $O(m^a)$ induced paths on $2a$ vertices and $O(nm^a)$ on $2a + 1$ vertices, and it requires $O(k)$ time to detect whether a vertex extends a P_{k-1} into a P_k, we have the following corollary:

Corollary 3. *Let $G = (V, E)$ be a connected graph with $|V| = n$ and $|E| = m$, and let $k \geq 5$ be a constant. By implicitly running DFS on $G^{(k-1)}$ it is possible to detect whether G contains a quasi-hole on at least k vertices in $O(n^2 m^{p-1})$ time when $k = 2p$, and in $O(n^2 + nm^p)$ time when $k = 2p + 1$.*

The space required is $O(m^{p-1})$ when $k = 2p$, and $O(nm^{p-1})$ when $k = 2p + 1$. According to Lemma 4 and Corollary 3, we finally get the following result:

Theorem 9. *Let $G = (V, E)$ be a connected graph with $|V| = n$ and $|E| = m$. It is possible to recognize whether $G \in sDH(2)$ in $O(n^2 m^2)$ time and $O(m^2)$ space.*

6. Conclusions

In this paper, we studied the class $sDH(2)$. It contains each graph G with stretch number less than two, that is $s(G) < 2$. These graphs form a superclass of the well studied distance-hereditary graphs, which corresponds to graphs with stretch number equal to one.

For the class $sDH(2)$ we provided: (1) a characterization based on listing all the minimal forbidden subgraphs, (2) a characterization based on cycle-chord properties, and (3) a recognition algorithm that works in $O(n^2 m^2)$ time and $O(m^2)$ space. This algorithm exploits the result based on the cycle-chord property to detects quasi-holes in a graph; it is a simple adaptation of the algorithm provided in [15] for detecting holes.

The characterizations found seem to suggest that the graphs in $sDH(2)$ and those in $DH(1)$ may be really similar in structure and hence properties. As a consequence, it would be interesting to determine whether the class $sDH(2)$ can be also characterized according to generative operations (we remind that the generative properties resulted as the most fruitful for devising efficient algorithms for distance-hereditary graphs). This problem has been partially addressed in [18,19].

On the contrary, Theorem 1 could suggest that graphs with stretch number greater or equal to two may have a completely different structure with respect to those in $DH(1)$.

Another possible extension of this work could be to investigate in the class $sDH(2)$ other specific combinatorial problems that have been solved in the class of distance-hereditary graphs.

Funding: This research received no external funding.

Institutional Review Board Statement: Not applicable.

Informed Consent Statement: Not applicable.

Data Availability Statement: Not applicable.

Conflicts of Interest: The author declares no conflicts of interest.

References

1. Howorka, E. Distance-hereditary graphs. *Q. J. Math.* **1977**, *28*, 417–420.
2. Bandelt, H.J.; Mulder, H.M. Distance-hereditary graphs. *J. Comb. Theory Ser. B* **1986**, *41*, 182–208.
3. Brandstädt, A.; Le, V.B.; Spinrad, J.P. *Graph Classes: A Survey*; Society for Industrial and Applied Mathematics: Philadelphia, PA, USA, 1999.
4. Brandstädt, A.; Dragan, F.F. A linear-time algorithm for connected r-domination and Steiner tree on distance-hereditary graphs. *Networks* **1998**, *31*, 177–182.
5. Chang, M.S.; Hsieh, S.Y.; Chen, G.H. Dynamic Programming on Distance-Hereditary Graphs. In *International Symposium on Algorithms and Computation*; Lecture Notes in Computer Science; Springer: Berlin/Heidelberg, Germany, 1997; Volume 1350, pp. 344–353.
6. Gioan, E.; Paul, C. Dynamic distance hereditary graphs using split decomposition. In *International Symposium on Algorithms and Computation*; Lecture Notes in Computer Science; Springer: Berlin/Heidelberg, Germany, 2007; Volume 4835, pp. 41–51.
7. Lin, C.; Ku, K.; Hsu, C. Paired-Domination Problem on Distance-Hereditary Graphs. *Algorithmica* **2020**, *82*, 2809–2840, doi:10.1007/s00453-020-00705-7.
8. Nicolai, F.; Szymczak, T. Homogeneous sets and domination: A linear time algorithm for distance-hereditary graphs. *Networks* **2001**, *37*, 117–128.
9. Rao, M. Clique-width of graphs defined by one-vertex extensions. *Discret. Math.* **2008**, *308*, 6157–6165.
10. Cicerone, S.; Di Stefano, G.; Flammini, M. Compact-Port Routing Models and Applications to Distance-Hereditary Graphs. *J. Parallel Distrib. Comput.* **2001**, *61*, 1472–1488, doi:10.1006/jpdc.2001.1728.
11. Esfahanian, A.H.; Oellermann, O.R. Distance-hereditary graphs and multidestination message-routing in multicomputers. *J. Comb. Math. Comb. Comput.* **1993**, *13*, 213–222.
12. Cicerone, S.; Di Stefano, G. Graphs with bounded induced distance. *Discret. Appl. Math.* **2001**, *108*, 3–21, doi:10.1016/S0166-218X(00)00227-4.
13. Cicerone, S. Characterizations of Graphs with Stretch Number less than 2. *Electron. Notes Discret. Math.* **2011**, *37*, 375–380, doi:10.1016/j.endm.2011.05.064.
14. Cicerone, S.; Di Stefano, G. Networks with small stretch number. *J. Discret. Algorithms* **2004**, *2*, 383–405, doi:10.1016/j.jda.2004.04.002.
15. Nikolopoulos, S.D.; Palios, L. Detecting Holes and Antiholes in Graphs. *Algorithmica* **2007**, *47*, 119–138, doi:10.1007/s00453-006-1225-y.
16. Hammer, P.L.; Maffray, F. Completely separable graphs. *Discret. Appl. Math.* **1990**, *27*, 85–99.
17. Cormen, T.H.; Leiserson, C.E.; Rivest, R.L.; Stein, C. *Introduction to Algorithms*, 2nd ed.; The MIT Press and McGraw-Hill Book Company: New York, NY, USA, 2001.
18. Cicerone, S. Using Split Composition to Extend Distance-Hereditary Graphs in a Generative Way—(Extended Abstract). In *International Conference on Theory and Applications of Models of Computation*; Lecture Notes in Computer Science; Springer: Berlin/Heidelberg, Germany, 2011; Volume 6648, pp. 286–297; doi:10.1007/978-3-642-20877-5_29.
19. Cicerone, S. On Building Networks with Limited Stretch Factor. In *Web, Artificial Intelligence and Network Applications, Proceedings of the Workshops of the 34th International Conference on Advanced Information Networking and Applications, AINA*, Caserta, Italy, 15–17 Aprli 2020 ; Advances in Intelligent Systems and Computing; Springer: Berlin/Heidelberg, Germany, 2020; Volume 1150, pp. 926–936; doi:10.1007/978-3-030-44038-1_84.

MDPI

St. Alban-Anlage 66

4052 Basel

Switzerland

Tel. +41 61 683 77 34

Fax +41 61 302 89 18

www.mdpi.com

Algorithms Editorial Office

E-mail: algorithms@mdpi.com

www.mdpi.com/journal/algorithms

www.ingramcontent.com/pod-product-compliance
Lightning Source LLC
LaVergne TN
LVHW070546100526
838202LV00012B/392